# Introduction to
# Travel Journalism

Lee B. Becker

GENERAL EDITOR

Vol. 5

The Mass Communication and Journalism series
is part of the Peter Lang Media and Communication list.
Every volume is peer reviewed and meets
the highest quality standards for content and production.

PETER LANG
New York • Washington, D.C./Baltimore • Bern
Frankfurt • Berlin • Brussels • Vienna • Oxford

John F. Greenman

# Introduction to Travel Journalism

## On the Road with Serious Intent

Foreword by John Maxwell Hamilton

PETER LANG
New York • Washington, D.C./Baltimore • Bern
Frankfurt • Berlin • Brussels • Vienna • Oxford

Library of Congress Cataloging-in-Publication Data
Greenman, John F.
Introduction to travel journalism: on the road with serious intent /
John F. Greenman.
p. cm. — (Mass communication and journalism; v. 5)
Includes bibliographical references and index.
1. Travel journalism. I. Title.
PN4784.T73G74    070.4'491—dc23    2012011109
ISBN 978-1-4331-1420-5 (hardcover)
ISBN 978-1-4331-1419-9 (paperback)
ISBN 978-1-4539-0834-1 (e-book)
ISSN 2153-2761

Bibliographic information published by **Die Deutsche Nationalbibliothek**
**Die Deutsche Nationalbibliothek** lists this publication in the "Deutsche
Nationalbibliografie"; detailed bibliographic data is available
on the Internet at http://dnb.d-nb.de/.

© 2012 John F. Greenman
Peter Lang Publishing, Inc., New York
29 Broadway, 18th floor, New York, NY 10006
www.peterlang.com

Printed in the United States of America

*For Alice*

# CONTENTS

# ACKNOWLEDGMENTS

I wish to acknowledge the contributions others made to this text.

Series Editor Lee B. Becker encouraged me to write this text; were it not for his encouragement, I wouldn't have. Professor Becker, who is my colleague in the University of Georgia's Grady College of Journalism and Mass Communication, recognized my commitment to a more journalistic approach to travel writing and thought there was an audience for this book wider than just my students. We both hope he was right.

Brian Creech, a doctoral student and my graduate research assistant in 2010–2011, read and outlined the literature that underpins much of the academic content in this text. Brian also contributed insights from his time as program assistant for the Travel Writing in Cambodia study abroad program that I direct.

Borrell Associates provided estimates and forecasts of the U.S. local advertising and promotion spending that buttresses the advertising analysis in Chapter 7. My gratitude to Colby Atwood, president of Borrell Associates, and Kip Cassino, a former colleague of mine at Knight-Ridder, who developed the model on which the estimates and forecasts are based.

Aaron Marshburn, one of the brightest and most intrepid undergraduate students I've known, tested some of the ideas in this text while traveling in Asia and Africa. As importantly, Aaron contributed his "Making Friends Model" for inexpensive, effective travel that is included in Resources.

Travel journalist Mary Bergin shared her week-long experience at the Poynter Institute learning to be a better entrepreneur for the case study "The Diary of an Entrepreneurial Travel Journalist in Training." Jane Blunschi recounted her experience with the crowd-funding site, KickStarter, for the case study "Crowd Funding a Travel Book about Elvis Presley's Birthplace."

Anettra Mapp, the administrative specialist I share, gathered data about leading newspaper travel sections.

Carolyn McKenzie and Don E. Carter endowed the distinguished professorship in journalism I hold at Georgia. Their support helped make this work possible.

My wife, Mary Alice Budge, professor emeritus of English at Youngstown State University, read the manuscript at several key points and offered questions, editing suggestions and encouragement.

Thanks to all.

**Permissions:**

Austin, Cynthia J. High population growth rates discourage education, foster poverty. Athens: Grady Journal, copyright Cynthia J. Austin, 2010. Reprinted by permission of the author. All rights reserved.

Bergin, Mary. www.roadstraveled.com. Copyright Mary Bergin, 2011. Reprinted by permission of the author. All rights reserved.

Greenman, John F. Caution: Black flies www.ecomonteverde.com. Copyright John F. Greenman. Reprinted by permission of the author. All rights reserved.

Greenman, John F. Tips and cautions. St. Petersburg: The Poynter Institute's NewsU. Copyright John F. Greenman, 2008. Reprinted by permission of the author. All rights reserved.

Marshburn, Aaron. The Making Friends Model. Unpublished manuscript. Copyright Aaron Marshburn, 2011. Reprinted by permission of the author. All rights reserved.

Randall, Brianna. When it comes to small-business loans, women are preferred. Athens: Grady Journal. Copyright Brianna Randall, 2010. Reprinted by permission of the author. All rights reserved.

Swick, Thomas. The 9 Elements Missing from the Conventional Travel Story. Tip sheet distributed at the 2009 Nieman Conference on Narrative Writing. Copyright Thomas Swick, 2009. Reprinted by permission of the author. All rights reserved.

Wilson, Elizabeth W. Choosing to become a monk for religious—and practical—reasons. Athens: Grady Journal. Copyright Elizabeth W. Wilson, 2010. Reprinted by permission of the author. All rights reserved.

# FOREWORD

About twenty years ago, I set out to repeat a pair of Panama-based adventures by Richard Halliburton, one of the 20th century's most picturesque travel writers. In his own time, Halliburton was ridiculed by sophisticated readers for his romantic, perpetual-college-boy perspective on the world. In our time, his writing seems mostly fit for children. Yet, he is useful signpost in the development of travel and travel writing.

In the 1920s and 1930s, when the world became a playground for Halliburton, travel changed. No longer was travel mostly for the leisure class or explorers who might very well not live to make the return passage. Between the two world wars, travel became an industry that offered affordable tickets for the average wage-earner. Droves of college students went on foreign excursions, as Halliburton did when he graduated from Princeton in 1921.

Halliburton was not merely an expression of this trend. His books kindled the desire for travel among the great mass of Americans. He described a world open for general admission, yet still offering adventure, albeit adventure in which hardships consisted of little more than late trains and diarrhea. His books were best sellers. Handsome and not embarrassed about wearing capes, Halliburton had the tools for celebrity. Ladies swooned when he swept into lecture halls. Youngsters read *The Royal Road to Romance* and his other books and decided they wanted to be Halliburton themselves. One of these was Edgar Snow. Destined to be one of the greatest foreign correspondents of his generation, Snow started out on his first foreign trip with the simple idea of just travel writing. "Happiness . . . meant but one thing," he told his parents in a letter written in Halliburton's style. "And that was travel!! Adventure! Experience!"

My quest in Panama was to find out if it was possible to do what Halliburton had done there in 1928. How much had travel changed, I wondered?

The first adventure was Halliburton's famous swim of the Panama Canal. He traversed the entire 50-mile stretch, going from the Gulf of Mexico to the Pacific Ocean. The feat required him to get permission to pass through the locks at each end, just as a ship would do, something that had never been asked for before and, as I discovered, will never be given again. When I said I wanted to swim the canal the same way, public relations staff directed me to a visitors' center where I was shown an old movie clip of Halliburton messing around in the locks. The best I could do was quietly rent a boat and swim in Gatún Lake, which makes up part of the canal on the east side.

Halliburton's second adventure was to climb Cerro Pirre, a mountain along the border with Columbia noted for its abundance of flora and fauna–and snakes. In typical cavalier fashion, he bought into the ill-informed suggestion from some local that it was from this vantage point that Vasco Núñez de Balboa first spied the Pacific. With the help of an environmental group, I put together a small party guided by an Emberá Indian. He was seconded by an individual of uncertain pedigree who energetically swung his machete as we climbed, shouting "Te quiero, carajo!"—"I love you, damn it!" After several days, we reached the top, which is today marked by a geodetic marker. Judging from a guest book at a camp at the base of the mountain, we were the first to do so in several years. Testing my trip against Halliburton's account, I concluded that he did not reach the summit, but probably a high point elsewhere. In any case, from no vantage point, including the actual summit, can a traveler see the Pacific. The canopy is too thick.

All the restrictions on the canal today illustrate how royal roads of romance have become paved and dotted with stop signs. The mountain climb showed there are still interesting detours that need to be protected and cherished. Both offer stories that are worth reporting, but rarely get the attention they deserve.

This is why John Greenman's book is needed. As the world has become more organized for the tourist, much travel writing has tended toward the sort that appears in Sunday newspapers. Those sections exist not because newspapers care deeply about the subject. They exist because travel agents, airlines, and resorts want a congenial place to advertise their services. Sunday travel stories describe pleasant passages on luxury liners through the Panama Canal. They have no interest in Cerro Pirre.

This is a how-to book that does not seek the lowest common denominator found in those Sunday sections. To be sure, it offers sound practical suggestions on how to write well and get published. But there is much more. It stretches the idea of travel writing into what Greenman calls "travel journalism." Travel journalism leads to better coverage of advertisers in travel sections, showing how the travel industry in its zeal to make destinations comfortable deadens them. It goes beyond the always rosy view that Halliburton had of the places he visited. Where he only saw fun, travel journalists show us poverty and disease. This sort of journalism helps us be responsible tourists. It helps us, even, learn to travel. With all the efforts made to ensure that tourists are not inconvenienced, we sometimes forget that the idea of travel is to experience that which is new and, thus, not comfortable. (One rule in this regard is always order food that you never dreamed of eating. I concede the sliced snakes heads and leeks in Guangzhou were ghastly; but the horse tartare in Lucca, Italy, was one of the best dishes I have ever eaten.)

As Greenman tells us, travel journalism can lead to plain old journalism. Again Halliburton is illustrative. Try as he did from time to time to become a serious writer, he was trapped in his own persona. He remained a jejune adventurer until he died in 1939 when a junk he was sailing across the Pacific Ocean was lost in a storm. But Halliburton's sans souci did not prevail for many of those who initially went abroad with the dream of emulating him. The diaspora of American youth who travelled in the 1920s and 1930s with an inclination toward writing became the backbone of the distinguished corps of correspondents who foresaw World War II and then covered it.

In an important way, Greenman takes us back to a time when neither travel nor travel writing was so commercialized. Not that money was not involved. Always writers wanted to sell their books and articles. But in the pre-Halliburton age, writers were not trying to sell a destination. They took readers by the hand into unknown places and let them experience it. Even the misanthropic writer Evelyn Waugh, whose travel books should really be called anti-travel books, brought foreign places to life.

Reading John Greenman's book I felt a bit as did when I climbed down from Cerro Pirre. On my way out of the jungle, we tramped to an Emberá village, a place of simple dwellings on stilts, bare-breasted woman doing their house work, and children playing. As night fell, I could see on one side a nearly full moon and on the other the last minutes of sun illuminating Pirre's long ridgeline. Having reached the top and now admiring it from afar, I rejoiced in the possibilities that still exist for experiencing the world if one is willing to go off the beaten path, even if that merely means looking for the byways of a European capital rather than signing up for potted tours.

Thanks to John Greenman for giving journalists a guide with wide horizons for serious reporting.

John Maxwell Hamilton

# INTRODUCTION

Some serious journalists like to belittle travel writers and their work. The discipline requires "no particular expertise." Its practitioners are "handmaidens" to the travel and tourism industry. They take "freebies" in return for favorable accounts of "must-visit" destinations.

Such belittling is valid in far too many cases. It shouldn't—and needn't—be this way. Travel and tourism is the world's largest commercial services industry.[1] Nearly a billion travelers arrive at international destinations annually. They travel for a widening set of reasons: business, leisure, adventure, study, volunteering. Others long to share their experiences.

Journalists should take seriously this large, fast-growing, diverse industry. Doing so now is especially important, as spending on foreign news coverage is in sharp decline.[2] Indeed, it makes sense to think of travel journalists as a type of independent foreign correspondent.

This text provides independent, public service–oriented journalists with the skills and knowledge they need to cover the travel and tourism industry, in order to provide travelers with credible news and information and to report significant trends and developments across the world.

The leading texts on travel writing presume the reader hopes to earn a living as a travel writer. A typical title: *How to Write—and Sell—Your Own*

*Travel Experiences*.[3] This text presumes the opposite: With rare exceptions, independent, ethical, substantial travel journalism is, at best, an adjunct to the regular, paying assignment of traditionally employed or freelance journalists. That said, there are entrepreneurial skills to be learned and practiced in order to produce and publish—and fund—travel journalism.

The audience for this text embraces traits common to journalism-school graduates: Their identity is independent. Their purpose is service to readers and viewers. Their method is informed but neutral.

Travel journalism is travel writing produced to a journalistic standard. This seems like a tautology, but it's not. Much travel writing is not produced to a journalistic standard. The coverage is so allied with the travel industry that it is more a product of public relations than journalism. The writers and their expenses are subsidized by the travel industry or by government travel offices and the subsidies are rarely disclosed to readers. Travel journalism is different.

Travel journalism is about travel just as business journalism is about business, health-care journalism about health care. Story types, topic niches and approaches vary: a news story about mega cruise ships damaging Venice's fragile infrastructure; an advice piece about apps that track airfares; a destination guide that explores a familiar place in an unfamiliar way; a journey along an ancient pilgrim's track.

What doesn't vary is the journalistic standard: pursuing stories according to relevance, usefulness and interest.

To succeed over time, basic multimedia skills are essential. Language and geography skills are helpful, but not required. Traits that will aid travel journalists include: wonder about places they've not visited; openness toward the ways of unfamiliar cultures; willingness to endure mild discomfort in order to travel frugally; and courage to find and tell unexpected stories. The most important personality traits are energy and intrepidness.

There are four common misconceptions and mistakes to be avoided. The first is a belief that others want to read or view "my travel experience." Second, a pack-journalism tendency to replicate the accepted view about well-traveled places. Third, an over-reliance on observation as a primary source for travel stories and, as a result, a neglect of all of the other sources common to journalism. And, fourth, the failure to include a so-called "nut graf" that tells the reader what the story is about and why the reader should care.

This text approaches travel journalism through a series of chapter-length questions and answers. By way of introduction, they are:

# Chapter 1—Who Is the Travel Journalist?

The travel journalist is a journalist first. Travel, broadly understood, is the subject matter of this journalist's work, just as business, education, or health might be the subject matter of another journalist's work. The travel journalist seeks to engage readers by making important news and information interesting—when, where, and how it's needed. The travel journalist's method is always journalistic: an informed neutral searching for the best available version of the truth.

# Chapter 2—Who Is the Audience for Travel Journalism?

The audience for travel journalism is large and well understood.

Tourism is one of the world's largest industries—some say the largest. Travel and tourism contribute about $6 trillion to the global economy, about 9 percent of the world's economic activity.[4]

We know a great deal about the demographics of these travelers. We know, for example, how old they are, whether they are married, have children, own or rent. We know how much they earn—by age, by family status, by household—and how they use credit cards to spend their earnings.

We also know a great deal about the psychographics, the personality traits, of these travelers.

Destinations hoping to lure these travelers use demographic and psychographic data in their marketing. Travel journalists, and the publications for which they write, do the same to target readers for their stories.

# Chapter 3—How Do You Reach the Audience for Travel Journalism?

Whether in print or online, travel journalism tends to fall within four story types: News, Service and Advice, Destination, and Journey. Chapter 3 defines the four types based on a consensus among practitioners, the dominant texts, and academic literature.

In brief, they are:

- News: Enterprise and investigative coverage of travel and tourism, one of the world's largest industries.

- Service and advice: Richly sourced, serving the traveler rather than the industry. Making use of increasingly sophisticated technology and application of social media.
- Destination: Nearby or at a distance, the place where you find the story, where you discover "sense of place."
- Journey: Getting to the place where you find the story.

The types range from the most "newsy," which is News, to the most "literary," which is Journey.

There are niches worth exploring within each story type. They help fulfill what *Conde Nast Traveler* says travelers are looking for—"authenticity, connection, and, most important, the chance to help (at least not to harm) the places that give us pleasure."[5] We'll explore more than a dozen niches.

## Chapter 4—What Are the Essential Skills and Knowledge for the Travel Journalist?

In Chapter 4, we move from concepts to skills, beginning with sourcing. It's often observed that journalists depend too little on observation—their senses of sight, smell, touch, taste, and hearing. The opposite could be said of travel writers: They depend too much on observation. The travel journalist must use all of the sourcing methods common to journalism: interviews, secondary sources, data, documents—as well as observation.

Topic influences—even dictates—story elements. But it is safe to say that regardless of topic, these elements will appear in some form and place in all travel stories: a lead, followed by a "nut graf," then background and context, the body, and an ending.

"All good stories have a structure," says James Stewart, "which unifies even seeming disparate elements."[6] The story structure answers the writer's question: Where do I go next? And the story structure reassures readers that there is a "logic" to the story. Chapter 4 examines several structures typical of travel journalism.

## Chapter 5—How Is Getting "Published" No Longer Just about Query Letters to Newspaper and Magazine Editors?

A lot is known about getting published in print, from the dominant "how-to" books, and from newspaper and magazine "submission guidelines," which are better. The *Travel Writer's Handbook*, now in its sixth edition, devotes 28 pages of instruction to writing query letters to newspaper and magazine travel editors. The number of pages it devotes to online outlets? Two.[7] Yet, newspaper Travel sections and travel-related magazines are in decline, and online media is the fastest-growing source of travel news and information. Chapter 5 explores the opportunity of getting published by online travel sites owned by others.

# Chapter 6—What Are the Opportunities for the Entrepreneurial Travel Journalist?

The opportunity, of course, is do it yourself. Become an entrepreneur.

Though historically reluctant, and often ineffective, all manner of journalists are trying entrepreneurship—some by choice, others by necessity. Their starting point may be a "boot camp" offered by a group such as We Media, or a course in a journalism graduate school such as Arizona State University—or a seminar at Poynter.

"Dos and don'ts" lists are common on sites such as Knight Digital Media Center.

Underpinning the work is a recognition that the reader, not the journalist, is in charge and the entrepreneurial journalist will serve the reader.

Success is recognized by money and that other journalism staple—awards. Chapter 6 shows you how to get started.

# Chapter 7—What Are the Funding Opportunities for the Travel Journalist?

Many travel outlets won't use stories based on industry-subsidized travel, yet few outlets pay the journalist's travel expenses, and the fees they pay for the stories aren't enough to cover the cost of the travel. So the writer accepts industry-subsidized travel and publishes in lesser outlets.

The conflict of interest, apparent or real, is obvious. The resolution isn't.

Some say disclosure resolves the conflict: Tell the reader who paid for the travel and let the reader decide. But my research indicates that disclosure, at least in the U.S., is unusual.

Another resolution is explored in Chapter 7: nontraditional funding. Early evidence indicates that as many as four sources of nontraditional funding are being exploited by travel journalists—especially entrepreneurial travel journalists—who choose not to be compromised by industry-subsidized funding: self-funding, grant funding, crowd funding, and investor funding.

As we delve more deeply into these questions and answers, you'll hear the voice of travel journalists. Nearly 200 writers, photographers, producers, and editors answered questions posed in online surveys.[8]

Each chapter offers opportunities to test your understanding of the material and to practice the skills each chapter teaches. Completing all of the assignments will provide you with a starting portfolio of travel journalism.

Five case studies are also included, offering in-depth stories about key issues. They are:

- Case study 1—Undergraduate journalism students travel to Siem Reap, Cambodia, to test a more journalistic approach to travel writing. Judge whether their work is independent, ethical, and substantial.
- Case study 2—A convention and visitors bureau tours a group of travel writers through a Civil War venue in Corinth, Mississippi. Examine whether subsidized travel influenced their work.
- Case study 3—A step-by-step guide to constructing a travel-related website. Decide whether this form of travel journalism is consistent with your ambition and skills.
- Case study 4—The diary of an entrepreneurial travel journalist in training. Follow the footsteps of a traditional travel journalist making the leap toward entrepreneurship.
- Case study 5—Crowd funding a travel book about Elvis Presley's birthplace. Consider the pros and cons of offering rewards to strangers who fund a travel journalism startup.

The text concludes with Resources tailored to the travel journalist. No claim is made about comprehensiveness; rather, these are selections based on credibility, access, and cost.

At bottom, this text argues that travel writing and travel journalism are not the same, that travel writers and travel journalists are not the same. They differ in identity, purpose, and method. The travel writer looks in a mirror, tending toward memoir and autobiography. The travel journalist looks through a window: What's outside is what matters. The travel writer depends on and serves the travel industry. The travel journalist is independent of the travel industry and serves the public. The travel writer is subsidized. The travel journalist pays his own way.

It is left to the reader to decide whether a more journalistic approach to travel writing is possible.

# · 1 ·

# WHO IS THE TRAVEL JOURNALIST?

The travel journalist is a journalist first.

J. Douglas Tarpley writes that journalism's myriad codes of ethics reveal something of a "canon" of journalism.[1] Journalists, these codes reveal, hold three things to be "important, desirable, and worthwhile."[2] They are:

- Journalism gives readers, listeners, or viewers two dimensions of information—the facts and an understanding or "interpretation" of these facts.
- Information is published to help the audience make intelligent and informed decisions as citizens as well as in their personal lives.[3]
- Journalism serves as a marketplace for the distribution of goods and services.

Similarly, Tarpley found, the codes reveal "principles" on which journalists stand. They include a commitment to voluntary guidelines, truth telling, independence, and a concern for appearances. The codes also reveal "standards" of practice, including accuracy, impartiality, fairness, and a commitment to challenge government and "other social institutions engaged in the public's business."[4]

Others have looked at these codes and drawn the same conclusions.[5] But no one doing this work has so captured the attention of journalists as Bill Kovach and Tom Rosenstiel. Their *Elements of Journalism* (2001) identifies a set of rules based on such canons, principles, and standards.

Kovach and Rosenstiel set out to "engage journalists and the public in a careful examination of what journalism was supposed to be."[6] They conducted 21 public forums attended by 3,000 people. They interviewed 100 journalists. They researched content and history.

The slim volume that emerged—discussing just 10 "elements" over 199 pages—became journalism's de facto standard of best practice. It has been widely adopted by journalism schools, and in 2010 was ranked eighth in sales by Amazon among journalism textbooks. No other "theory of journalism" text ranks as high. It is widely cited by journalism scholars, with some 287 citations by end 2011, according to Google Scholar.[7] "An immensely valuable primer," according to the late journalist David Halberstam, "on who we are, what we do, and how we should do it."[8]

Many of the "elements" are relevant to the travel journalist. They are the elements of independence, ethics, timeliness, substantiality, and a public-service orientation. When they are evident in the work, they differentiate the travel journalist and travel journalism from the travel writer and travel writing.

Yet, and here is the hard part: The travel journalist must adapt to what Folker Hanusch calls the "special issues, exigencies" of travel journalism.[9] According to Hanusch and others, these include paying for the cost of travel, market orientation, and representing foreign cultures. Handled wrong, each one challenges the idea that the travel journalist is a journalist first.

Here's a preliminary look at each one:

Paying for the cost of travel: Travel, especially foreign travel, is expensive. The question is, who pays? Is it the travel journalist, expecting to recoup the cost upon publication? Or is it the travel publication, upon assigning or accepting the work? Either answer is "correct" in terms of standards of independent, ethical practice. But neither answer squares with this reality of travel journalism: Virtually <u>no</u> publication pays the travel costs of the travel journalist. Enter the travel industry, fully willing to pay the travel costs of the travel journalist in return for something it cannot buy: highly influential editorial content.

Market orientation: The question is, what market does the travel journalist serve? Does it serve the market for news and information about one of the world's largest, most complex industries? Or, does it serve the market for entertainment about journeys to "hidden" . . . "secret" . . . "unexplored" destinations. Thomas Hanitzsch calls the former, "journalism in the public interest," the latter, "journalism that addresses audiences as consumers."[10] The former signals low market orientation, the latter, high market orientation. For the travel journalist, it cannot be either/or, but both.

Representing foreign cultures: The question is, how does the travel journalist, when reporting outside her own culture, portray what cultural studies experts call "the other"? One answer is to treat them journalistically, gathering and verifying information, using all of the methods available to journalists: interviews, secondary sources, data, documents, and observation. Oddly enough, this rarely occurs: Carla Santos found that just one-ninth of the travel articles she studied referred to any communication with local residents of the destination.[11] The consequence of this, Santos says, is that "American readers are provided with representations that serve to reconfirm their own values and beliefs based on marketing strategies aimed at creating an interesting destination and culture for Americans to visit."[12] Hardly journalistic. Perhaps this explains this exasperated travel editor's injunction to writers: Talk with "at least one local, won't you."[13]

So how is the travel journalist expected to adapt? Were there easy answers, no one would categorize them as the "special issues, exigencies" of the travel journalist. Each will be considered further in succeeding chapters.

## Purpose

The travel journalist seeks to engage readers by making important news and information interesting—when, where, and how it's needed.

Certainly, the travel industry generates lots of important news and information. First, it is newsworthy because of its size. Travel is a global industry, large and growing, employing hundreds of millions. Second, it is newsworthy because of government's role. Government regulates many aspects of the travel industry, including access to and the safety and health of transportation, housing, and food. Government also taxes travel at every level of consumption. And, third, travel is newsworthy because all manner of man-made and natural disasters impact travel.

Here are representative examples of story topics, drawn from the news archives of the U.S. Travel Association over a recent 15-month period:[14]

- The impact on tourism from disasters such as oil spills, hurricanes, and volcanic ash; from the world economic crisis; from diseases such as swine flu; from terrorism; and from rising gasoline prices

- Travel boycotts inspired by government actions, such as the Arizona immigration law
- The travel industry's political clout as seen in passage of the federal Travel Promotion Act, and in higher levels of government spending to boost travel, both within the U.S. and abroad
- Scrutiny of corporate and government travel "junkets"
- Airport security, practices of the Transportation Safety Administration, and air traffic control safety
- Jittery valuations of publicly owned travel companies
- Visa waivers

The travel journalist engages the reader by making news and information interesting, whatever the story type. In-depth news, enterprise, and investigative reporting. Authoritative, credible service and advice. Destination guides that reveal sense of place. Heroic journeys.

Again, whatever the story type, the purpose is journalistic: independent, ethical, timely, substantial, and public-service oriented.

# Method

The travel journalist's method is always journalistic. There are many possible definitions of journalistic method, but I like this one from a British editor whose name I long ago lost: an informed neutral searching for the best available version of the truth.

This is an artful definition of journalistic method, superior to the alternatives, including objectivity. It is particularly suited to travel journalism. The "informed neutral" is a specialist, but is impartial. "Searching for" is about discovery and verification. The "best available version of the truth" reminds us that journalism is both timely and tentative, humbled by the compromise of deadlines.

Seven methodological approaches that are common to journalism are particularly relevant to travel journalism. They are:

- Reporting travel industry news, judging news value according to the standard criteria of relevance, usefulness, and interest.[15]
- Investigating the travel industry. The archives of Investigative Reporters and Editors demonstrate that the travel industry is a rich lode for investigative journalism. Among its recent topics, in order of

frequency: tourist safety and health, sex tourism, defrauding of tourists, environmental damage, and government tourism organizations.[16]

- The "immersion" approach associated with travel journalists such as Vivian Gornick, who immersed herself in the middle-class life of Cairo for her book *In Search of Ali Mahmoud* (1973). "Instead of analyzing my subject," she says, "I merged with my subject."[17]
- In contrast is the "parachuting" approach, more often associated with foreign news reporting, but actually the most common approach of the travel journalist: Preparation here, trip there, writing here—what Ulf Hannerz calls "quick passages along the news landscape."[18]
- The "Tapping into Civic Life" approach identified with civic journalism promises what travel journalists and their audience crave: authenticity. "A window opens and you will see what really makes communities tick," advocates of this approach promise. You discover "the underlying forces and trends that shift and evolve; people's concerns as they struggle to make sense of things around them; people's hopes and aspirations and how they connect their lives to the larger world."[19]
- Review and criticism, similar to the work of drama, film, and book critics.

Thomas Swick noted a "change of direction, a refinement of method," as he examined the work of a "new generation of travel writers" in 2001. Some seek "more remote outposts," Swick noted, but others mine the domestic "to discover not the places people traditionally travel to but those they travel from." The old role of gathering images is no longer enough. "It must go beyond pictures," Swick says, "and root out meaning." Some even explore current events—the travel journalist as eyewitness.[20]

## Another Approach to Identity, Purpose, and Method

There is another way to think of the identity, purpose, and method of the travel journalist. That is for the travel journalist to reinvigorate, perhaps by redefining, a time-tested concept: the independent foreign correspondent.

Traditional approaches to foreign news are in sharp decline. Consider:

- Eighteen newspapers have closed every one of their foreign bureaus over the last 12 years.[21]

- Sixty-four percent of newspapers have reduced the amount of space apportioned to foreign news (as little as 2 percent today versus 10 percent 40 years ago[22]), and 46 percent have allocated fewer resources to cover foreign news.[23]
- The major U.S. television networks—ABC, CBS, and NBC—aired 2,070 minutes of foreign news in 2010, nearly 60 percent less time than two decades earlier when the Soviet Union was collapsing.[24]

Foreign news is "rapidly losing ground at rates greater than any other topic area," concludes a Project for Excellence in Journalism report, "The Changing Newsroom."[25] Yet, this ground is lost at a time when foreign news is as important as ever. "In a world that's more interconnected than ever," says John Walcott, "foreign reporting is an essential component of a complete news report."[26] And the audience for news remains interested. "No matter what year or what market," says Virginia Fielder, roughly six in 10 readers are "highly interested" in foreign news.[27]

One approach is to think of foreign news as local news. Jack Hamilton first suggested it in 1988. Hamilton's *Main Street America and the Third World* argued that "events overseas have an impact on the average American" and should be covered as a local story.[28]

Examples from Hamilton's work 25 years ago are still relevant today:

- A manufacturer of tobacco-products machinery turns to new markets in Asia after growth in the U.S. market stalls
- A retailer deconstructs the origin of the shoes on her shelves: U.S. leather, tanned in Asia, sewn in Brazil
- A U.S. defense contractor recognizes that rare minerals—cobalt, chromium, manganese, and platinum—essential in the production of key alloys come from China, India, the Philippines, and Zaire. More recently, the term applied to this approach is to *glocalize* the news.[29]

"Following chicken parts from an Alabama processing plant to Russian dinner tables" is an example of the idea. So is "going home to Mexico with a Missouri popcorn plant worker," and "linking Kansas job losses to Italian technology."[30]

Nearly 70 percent of regular readers "will read more non-local stories if they see how these stories affect them," according to a study for the Newspaper Association of America. "Straight reporting without the tie in won't help."[31]

A primer, "Bringing the World Home," produced by two journalism organizations, offers 19 pages of story ideas and story sources.[32] Poynter's NewsU offers a self-directed online course, Reporting Global Issues Locally,[33] which includes a simple tool to assess the opportunity to report global issues in your market. The International Reporting Project maintains an archive of such stories, too.[34]

Some of these stories can be reported from home. Others require travel.

The question is, might travel journalists, focusing more on newsy venues and angles, do this work? Were they to do so, they might reinvigorate, and perhaps redefine, the concept of the independent foreign correspondent.

The concept of the independent foreign correspondent is not new. Hamilton's most recent book, *Journalism's Roving Eye* (2009), traces the concept of the independent foreign correspondent from the beginning of the twentieth century. One "planned only a vacation jaunt." Another "decided I had to see the world." A third "went to Paris for two weeks and stayed five years."[35] This was the Roaring Twenties, Hamilton writes, when there was a sense that "another job and more money awaited their return." The period between the world wars "was something of a golden age" for the independent foreign correspondent. The news was "momentous," the outlets "plentiful," and Americans were "well liked."[36] After World War II, Americans "still struck out on their own to go abroad and be correspondents." Jonathan Schell went to Vietnam under the "preposterous claim" that he was a correspondent for his college newspaper, *The Harvard Crimson*.[37] By the end of the last century, economics and the corporatization of news worked against the independent foreign correspondent. The cost of living abroad was rising; the value of the dollar was declining. Hamilton quotes *USA Today*'s foreign editor, Elisa Tinsley, that there "is less tolerance for the kind of eccentricities" exhibited by independent foreign correspondents.[38]

How might now be a better time?

Technology is more favorable in several ways. First is the technology that facilitates independence. Handheld digital cameras replace film crews. Laptops replace editing tables. Satellite phones replace satellite trucks. Second is the technology that facilitates fast, cheap, online publishing of words, audio, and video. Third is the technology that lowers costs. Equipping an independent, multimedia, foreign correspondent costs as little as $10,000.[39]

A supportive infrastructure is emerging. A mix of start-ups are providing alternatives for independent foreign correspondents. One is the nonprofit Pulitzer Center on Crisis Reporting organized by Jon Sawyer.[40] Another is the International Reporting Project (IRP) developed by John Schidlovsky.[41] A third is GlobalPost, the effort of Philip Balboni.[42]

And, there's help with the cost of travel. Schidlovsky's IRP, for example, provides 40, five-week travel grants each year.[43] A dozen other organizations, in one form or another, do the same.[44] These grants, scholarships, and fellowships don't come with the conflicts of interest that are common in subsidized travel journalism.

A handful of travel journalists are leading the way. Ted Rall and Rory Stewart are often mentioned.[45] Stewart wrote *The Places in Between* (2006), an account of his 32-day walk from Herat to Kabul, Afghanistan, just after the Taliban fell in 2002. Similarly, Rall wrote *To Afghanistan and Back* (2002), based on his coverage of the war in the months preceding the Taliban's collapse. But neither Stewart nor Rall identifies himself as a travel journalist, and their book-length reports are an unrealistic model of this idea.

More realistic models are to be found elsewhere:

- Jeff Greenwald explores the tension between tourism and political, humanitarian, and environmental issues. His article "Bedeviled Island," which told the story of a Tasmanian cattle rancher who converted his land to a wildlife preserve, won a 2009–2010 Lowell Thomas Award. Greenwald also directs the organization Ethical Traveler, which encourages travelers to visit ethical destinations.[46]
- Daniel Brook takes readers on unexpected journeys. "The Architect of 9/11," a three-part series for the website Slate.com, traced the schooling, writings, and architectural career of the terrorist Mohamed Atta. Brook "invites us to look at a part of the world through the terrible eyes of hatred," wrote the judges who awarded Brook a 2009–2010 Lowell Thomas Award.[47]
- Joshua Hammer describes his work as "postcards from the edge," travel journalism about extremist ideology and religion, failed nations, foreign intervention, and individuals caught up in cataclysmic events.[48] A story for *Outside* about relief efforts in the war-torn nation of Chad, "Heartbreak, Chaos, Mayhem Hope?" won a 2009–2010 Lowell Thomas Award.

But not all travel journalists are like Greenwald, Brook, or Hammer. Some travel journalists will be unsuited for this work, just as no journalist is suited to every beat. But there may be a way to test suitability beyond the fundamental question, am I interested?

Stephen Hess surveyed both independent and staff foreign correspondents and measured the ways in which the former differed from the latter. Among the ways, the independent foreign correspondents are:

- Younger
- Less likely to be married
- More likely to be women
- Better educated and better off financially
- More proficient in languages[49]

Ulf Hannerz also has studied what kind of journalists are likely to be drawn to foreign correspondence. Perhaps like journalists generally, he writes, "they were likely to be people who were curious about things, liked easy contacts with others, would be quick to establish rapport and did not seek for depth in relationships."[50]

Recent experience with a bright, young undergraduate student convinces me that this work is within the reach of most travel journalists. His name is Aaron Marshburn and he took my Introduction to Travel Journalism class as an independent study. Base camp for Marshburn was Bangkok. From there, he found his way to Mae Sot, a trading center on the Thai border with Burma. Most foreign travelers pass through Mae Sot on the way to the Thee Lor Sue Waterfalls, one of the world's tallest. Not Marshburn. He was there to sneak his notebook and camera into a Karen refugee camp at Umpiem Mai, in the mountains south of Mae Sot. Among the refugees he met was Minnai, a soldier in the Karen National Liberation Army until he was blinded in a landmine explosion. "I would rather have freedom for my people than regain my sight," he told Marshburn.[51]

There is a long tradition of foreign correspondents doing travel journalism. CNN's Madrid bureau chief, Al Goodman, famously covers Spanish wines for *Wine Spectator*. When not covering Basque separatists, train bombings, or plane crashes for CNN, the magazine paid him to "dutifully sip [my] way through 30 Spanish wineries."[52]

The argument here is, let's flip it—let's have travel journalists doing independent foreign correspondence. Ready to get started? Training and advice are available. The most formal is offered by Transitions Online, a journalism training center in Prague. Its one-week Foreign Correspondent Training Course is taught by correspondents whose work appears in *The Economist*, the *New York Times*, and the *Christian Science Monitor*.[53] A self-directed online course, International Reporting Basics: What You Need to Know Before

You Go, is offered by Poynter's NewsU.[54] An excellent set of tips and cautions is in Goodman and Pollack's *The World on a String: How to Become a Freelance Foreign Correspondent* (1997).[55]

## Test Your Understanding

1. J. Douglas Tarpley writes that journalism's myriad codes of ethics reveal something of a "canon" of journalism. Journalists, the codes reveal, hold three things to be important, desirable, and worthwhile. What are they?
2. Many of the 10 "elements" in *The Elements of Journalism* are relevant to the travel journalist. Name five.
3. Folker Hanusch cites three issues he calls the "special issues, exigencies" of travel journalism. What are they?
4. Seven methodological approaches that are common to journalism are particularly relevant to the travel journalist. Identify and discuss three of them you regard as most relevant to your market or interests.
5. One idea is that the travel journalist should reinvigorate, perhaps by redefining, the concept of the independent foreign correspondent. Discuss the concept and its implications for travel journalism. Why might this be a good time to follow this approach?

## Practice Your Skills

Assignment #1: Assessing the opportunity to cover travel and tourism in your market. Answer these five questions about your market:

1. Is there a taxpayer-funded destination marketing association in your market? Often, they are called "conventions and visitor bureaus" and are funded by taxes on hotel and motel charges. If there is, how much revenue does it draw each year, and how is the money spent?
2. Is there a state tourism office in your state? If so, it has probably commissioned a third-party study on the statewide and county-by-county economic impact of tourism. Where does your local market rank? What explains your market's ranking?
3. Is there an important upcoming event or new attraction in your market? If so, what would a tourist need to know in order to fully benefit from attending or visiting? What if the tourist was value-oriented, traveling on a lean budget?

4. Is there a "place" in your market that could be described as underused, unexplored, or hidden? It could be an obscure neighborhood, a forgotten attraction, a "gem" known just to a few.

5. Can you imagine a half-day "journey" in your market that a tourist might want to take? Consider historic markers, charity thrift shops, the graves of storied characters, homes of a certain architecture, eco-friendly investments, and so forth.

## A Closer Look: Students Test a More Journalistic Approach to Travel Writing

Beth Pollak, an honors student majoring in biology and journalism, returned from a day-long reporting trip in rural Cambodia with a discovery.

"Ever hear of the sugar palm tree?" she asked. None of us had.

Pollak explained that the sugar palm is something of a wonder tree. Its root is used in traditional medicines said to treat illnesses ranging from malaria to STDs to measles. The wood is used in home construction and furniture. The leaves are used for thatching the roofs and walls of traditional Cambodian homes. And juice from the tree's fruit is processed into palm syrup and sugar. Palm sugar is a healthier alternative to its cane equivalent: It has a low glycemic index, meaning it is a slow-release energy source that won't produce the infamous sugar high. It is also full of vitamins and minerals, including calcium, potassium, and vitamin C.

"No wonder it's Cambodia's national tree," Pollak said. And an idea for a story.

Pollak was one of 10 University of Georgia students who joined me for the first year of Travel Writing in Cambodia, a laboratory-type program designed to test whether undergraduate students, in a month-long study abroad program, can demonstrate a more journalistic approach to travel writing.

"Eighty percent of travel writing is not journalistic," I told the students our first day in class. "Travel writers are too dependent on the travel industry, take money for travel and don't disclose it to readers, and write about too many familiar, upbeat, inconsequential topics."

We're going to model the other 20 percent: independent, ethical, substantial.

Pollak and her classmates got it immediately. Over the month we were in Cambodia, none proposed a survey of "the best backpacker hostels for college-age world travelers." Instead, they mined fresh, unexplored topics, born of the curiosity they exhibited on daily 3–5–hour reporting trips.

In a way, they knew no better. None had experience as travel writers. They were unaware that most travel writers follow different standards—compared with what they were learning—with respect to identity, purpose, and method. None had ties to the travel industry. They wouldn't dream of taking expense money from a source. And they couldn't imagine flying halfway around the world in search of inconsequential fluff.

"Think of yourselves as independent foreign correspondents," I said. They did.

Study abroad is an important early training opportunity for travel journalists. The cost is relatively low. Students experience a sustained, immersive period of study. They behave as a cohort, guided by a trustworthy, experienced teacher and travel journalist. We provide a safe, supportive, helpful environment so the students are free to study and work. In Cambodia, their assignment was a daily 3–5–hour reporting trip, sandwiched between a pair of 75-minute classes in 107-degree heat and the highest humidity of the year, on the brink of the monsoon season.

For me, as a journalism teacher, this was an opportunity to test ideas and approaches to teaching travel journalism. I wanted to see what would take, and what the class would resist.

Our venue was the province of Siem Reap, the highly touristed home of the Angkor temples, including Angkor Wat, the UNESCO World Heritage Site that is on every global tourist's must-see list. Siem Reap also is home to Tonle Sap Lake, a compelling and distinct environmental site. More than half of the province's 896,000 people live on less than $2 a day, yet entrepreneurship—fueled by a vibrant microfinance industry—flourishes. There is unexploded ordnance in the deep countryside, a leftover from the Khmer Rouge revolution of the 1970s. Monks train in monasteries. NGOs dig wells for clean water and teach English.

There are lots of travel journalism opportunities, but many have been raked over by seasoned travel journalists. The question was: Would these juniors and seniors, with little to call on beyond youthful curiosity and a couple of news- and feature-writing classes, find anything new?

Certainly Elizabeth Wilson did. A dual major in journalism and religious studies, Wilson focused on young monks. Typical travel portraits rarely get past description of the saffron robes, mostly because of the monks' culture: They are inaccessible, contemplative, mostly mute. But Wilson discovered an early-evening ritual called "monk chat." She interviewed dozens of monks over a week of evenings and came away with a discovery: Many poor, rural Cambodians enter the monastery not just for the spiritual calling, but for the education. Here's the nut graf Wilson wrote:

> In a country with high levels of poverty and illiteracy, monastic life turns out to be a pivotal tool in developing an educated class. In addition to strengthening their spirituality, young men with a hunger for education, but no opportunities to fulfill it, become monks to study.

> Much like young Americans who enlist in the military for the educational benefits, many young Cambodians choose to become Buddhist monks for the same reason. Becoming a Buddhist monk need not mean a vocation for the rest of one's life. Just as many American soldiers leave the military to go on to do other things, Buddhist monks may leave when they wish—and many do.[56]

Business student Brianna Randall, the only non-journalism major in the program, also found a fresh angle on a familiar story: the microfinance industry. Microfinance capitalizes small businesses formed by poor entrepreneurs. Although it's been around for 50 years, few outside the movement knew about microfinance until its founding thinker, Muhammad Yunus, won the 2006 Nobel Peace Prize. What Randall discovered is that in Siem Reap, almost every microfinance borrower is a woman. Here's how Randall introduced the discovery:

> About two minutes into a small village sits a house built of bamboo. A college-age loan officer named Lida Reach approaches the owner of a vegetable stand, a Khmer woman named Sopheaktra. In her hand is the weekly payment against her $250 loan, which she gives to Reach in exchange for a receipt.

> This scene repeats dozens of times each week in this historic Cambodian city, where women like Sopheaktra, who own the bulk of the country's small, often informal, businesses, are the microfinance industry's most-frequent, least-risk borrowers.

Then, Randall offered this as background, context, and a nut graf:

> Developed over the last 50 years, microfinance institutions have, according to global micro-lender Kiva, lent more than $2.5 billion to 16 million people worldwide. Cambodia is one of the most rapidly developing microfinance areas in the world according to the World Bank's International Finance Corporation. Here in Siem Reap, the micro-lending nonprofit Journeys Within Our Community (JWOC) focuses on women like Sopheaktra who are the "poorest of the poor" in Cambodia.

> Camilla McArthur, a JWOC managing director, said the organization prefers lending to female borrowers.

"Why?" I asked.

"Women are just better," McArthur said.

In fact, several studies on women's role in microfinance, including one by Allianz, conclude that women are more likely to repay their debt, cooperate more effectively, and make the micro-institutions themselves more efficient.[57]

Randall also is an example of a travel journalist exploiting all of the sourcing methods available to journalists. Within the first 200 words of her story—the equivalent of 4 standard column inches in print—Randall uses interviews, secondary sources, data, and documents, as well as that overused travel journalism standby, observation.

The students were encouraged to rely on curiosity as the key driver of story ideas. Elliot Ambrose was curious why the DVD of the movie he was watching, *The Killing Fields*, failed in the final minutes. Turns out it was a cheap, pirated copy, for sale in the Old Market downtown for about 25 cents. This led Ambrose to ask: What else is pirated, smuggled, and knocked off—and who buys this stuff? Others, of course, have asked this question, but Ambrose used these practices as a focused lens with which to observe Siem Reap's tourist-targeted retail economy.

The students were surprisingly resilient, even tough-minded. Cindy Austin talked her way into a maternity hospital in connection with her story about women's reproductive health. Working with a translator, she interviewed a dozen women on the first day of reporting. Pushing the translator to get what she needed meant being "approachable and fearless," Austin said, "but not rude or abrasive." From one of the interviews, Austin got this:

Some families don't see the correlation between more children and less money. They choose to have more children so that the older ones can help make money for the family and raise the younger children. Others, however, have caught on to the economic consequences.

"If you have kids, you are poor for a little while," says Liem, a woman in a squatter village here, as her third child sleeps at her feet. "If I have more kids I will be poor forever."

From another, she got this:

It seems that many women in the village, however, know about birth control and still choose to have a large number of children. Sok Pove, a mother of two boys in another village here, has a different motivation for having more kids.

"I'll keep having kids until I get a girl," Pove said.[58]

The students were also encouraged to experiment with different story structures appropriate to the type of story they were writing. For a story on how young students fund their education, Jayshri Patel adopted a structure associated with trend stories: anecdotal split lead, nut graf, background and context, back to the anecdote, then another and another. They were discouraged from the sort of breathless, follow-me-on-a-journey structures on which too many travel narratives rely. Yet, story-telling was central in their work, with lots of "show" as well as some "tell." Many of the students adopted what Roy Peter Clark calls a "broken line" narrative, where analysis follows action.

Colin Tom, for example, wrote about the surprising degree of prostitution in a venue known primarily for its temples. Tom led his story this way:

> Most travelers come here for the "divine inspiration" of the Angkor temples. Others come for a different inspiration.
>
> Atop the temples, sculptures of topless women—called apsara—were meant to appease 12th century kings and gods. The same sculptures adorn the walls of downtown nightspots like Temple Bar. But the purpose is different: to alert 21st century men to the sex trade.
>
> Prostitution here is as inescapable as the heat.[59]

Tom takes the reader into the action: "Temple Bar prostitutes gather by the pool tables or perch on furniture around the dancing areas. They laugh with one another in the back of the bar, periodically squeezing an arm or shooting a seductive glance at a tourist." Then Tom pulls back for an explanation: "Pub Street is a commercial strip for prostitutes who target travelers. Scantily clad women and tight-fitted transsexuals work the bars, lurk in alleys and doorways. Some walk the street. Cambodian men go elsewhere." Back and forth it goes, like a train racing down the track, Clark says, stopping occasionally at the stations.[60]

Students in this makeshift laboratory took away as much as I did. "Being curious connected us to the people we talked to, and I remember that," wrote Nicole Meadows a year after her experience. "That's the learning," she wrote. "Follow your curiosity."

# · 2 ·

# WHO IS THE AUDIENCE FOR TRAVEL JOURNALISM?

It is essential that travel journalists know their audience. Understanding the audience for travel journalism is essential for four reasons. First, the audience is large by almost any measure. In 2010 international tourist arrivals hit a record 935 million, with some 11 percent of them arriving in North America.[1] Second, the audience is varied. Whether viewed by "needs and desires" or "temperament and lifestyle," the audience is demonstrably diverse, as measured by demographics or psychographics.[2] Third, the audience is demanding: Sometimes it wants "advice, facts and knowledge" about how to travel;[3] other times it wants to get lost in someone else's adventure. And fourth, the audience is trusting. It perceives travel journalism to be "accurate"[4] and relies on travel journalism more than any other factor when making travel decisions.[5]

This is an audience, then, to be taken seriously. Yet audience plays a minor role in the books that dominate the "how-to" travel-writing market. Less than 2 percent of the pages of these books refer to "readers" or "audience."[6] Fortunately, information about the audience for travel journalism is well developed and widely available.

The audience for travel journalism is not one, but two, distinct audiences. One is the audience for leisure travel; the other is the audience for business travel. The audience for leisure travel no longer regards it as an expensive luxury enjoyed by the wealthy. Rather, leisure travel is a "psychological necessity" for all.[7] Leisure travel relieves the tension, stress, and pressure of everyday existence. It's a matter of "getting away from it all."[8] Business travel, on the other hand, tends to foster tension, stress, and pressure. So, psychologically, the audiences are different.

The audience for travel journalism is large and growing. The sizes of the audience and its economic impact usually are denominated in millions—even billions. Consider, for example, some U.S. data for 2009:[9]

- 1.9 billion trips[10] for leisure or business
- $704.4 billion spent by domestic and international visitors, excluding the cost of transportation
- $186.3 billion in wages paid to the 7.4 million workers directly employed in the travel business
- $113 billion in travel-related tax revenue paid to local state and federal governments

And that's just U.S. data—world data provides much larger figures.

The two best sources of travel-related data are the U.S. Travel Association for U.S. data and the United Nations World Tourism Organization (UNWTO) for world data. U.S. Travel is the travel industry's trade association. Although it is a nonprofit, some of its data is proprietary and expensive, but much of it, including the data cited here, is both nonproprietary and free. UNWTO data is typically made public through conference reports and press releases, but some of its data is available only to members.

The third thing we need to know about the audience for travel journalism is its composition. Demographics are one way to look at the composition of the audience for travel journalism. Psychographics are another way. The *Lifestyle Market Analyst* from Standard Rate and Data Service (SRDS)[11] is the single best source of information for both.

Here's what it contains and how you can use it.

Market Profiles: The U.S. is divided into 210 markets known to the marketing industry as DMAs, or designated market area. The profile of each DMA contains both demographic and psychographic information that can be used to understand the travel-related behavior of a given market. As an example, take the Charlottesville, Virginia, market. Here's what you can quickly ascertain from its market profile:

- Compared with national averages, Charlottesville is a well-educated, professionally employed, high-income market.

- Among the lifestyle choices of its households, three stand out: attending cultural and arts events, enjoying fine food and wine, and foreign travel. All three are significantly higher than national averages.
- An early spring travel story, anticipating European art, food, and wine festivals, would well serve the Charlottesville audience for travel journalism.

Lifestyle Profiles: Seventy-five lifestyle interests are analyzed. The analysis shows the demographics and psychographics associated with each lifestyle interest. Let's return to the Charlottesville example to see how this analysis helps.

We said that Charlottesville's market profile would justify an early spring story anticipating European art, food, and wine festivals. We said that, in part, because Charlottesville households are a third more likely to attend arts and cultural events than the average U.S. household. So, what else can we learn about households that attend arts and cultural events that might underpin the story?

We know from the Lifestyle Profiles that those households are more likely to be:

- Married, with two incomes and no children
- Health conscious, often cycling for transportation or exercise
- Tech-savvy, dependent on mobile wireless devices

So, slant the story towards married couples without children. Be sure the "if you go" boxes contain information on exercise centers, running trails, and bicycle rentals. Scope out the festival-related mobile apps. And so forth.

Two versions of the SRDS *Lifestyle Market Analyst* are available: One is an annually updated print version, likely to be found at the reference desk of any university library. The other is an online version, which is more powerful because of a mapping function. University libraries are likely to have access to the online version, too. If not, it's expensive: at least $725 per year.

The fourth thing we need to know about the audience for travel journalism is its behavior—the frequency, mode, and purpose of travel. How often do U.S. adults travel to domestic and foreign destinations? What is their mode of travel: air, ship, bus, auto? Why do they travel: business or leisure?

The best single source of this information is the periodic study "Travel Categories," prepared by Mediamark Research, Inc., for the Newspaper Association of America (NAA).[12] The most recent study was in 2008. NAA commissions the study to show advertisers how well newspapers reach the travel market, but you can use the underlying data about frequency, mode, and purpose of travel to better understand the audience for travel journalism.

Here's an example of what the data shows about domestic travel:

- 53.5 percent of U.S. adults traveled domestically at least once in the last 12 months, and 35.2 percent did so for a vacation.
- Nearly 40 percent of this travel occurred in the summer months—June, July, and August.
- Twice as many, 10.6 percent, used the Internet to book travel than used a travel agent, 5.0 percent.
- 21.9 percent flew on a scheduled airline and 17.5 percent are members of frequent flyer programs. Just 1 percent or fewer traveled by bus or rail.
- General sightseeing was the most popular activity, followed by shopping, visiting family and friends, going to the beach, attending a specific event, and visiting a national park.
- More backpacked and hiked, 2.6 percent, than played golf, 1.9 percent. More fished, 2.6 percent, than skied, 0.8 percent.
- The South is by far the most popular region to visit, 27.9 percent, followed by the West, 18.1 percent, the North Central, 13.7 percent, and the Northeast, 11.9 percent.
- 13 percent spent less than $1,000. Just 1.7 percent spent more than $5,000.

These data tell you how U.S. adults, as a whole, travel domestically. It's good data, from a credible source, that answers a lot of questions about the audience for travel journalism: These are the key trends. But these are national data. If your focus is a local market, how do you know whether the trend applies?

Two ways to check.

One way is to compare the demographics of the local market with the demographics of the U.S. The more they are alike, the more the NAA "Travel Categories" data apply.

Another way to check is by looking at the "index" numbers in the SRDS Market Profile data discussed above. If the index numbers in comparable categories are at or near 100, it's fair to conclude that the local market is like the U.S. as a whole.

Whichever way you employ, use caution.

That said, it may make for a better story if the local market is different from the U.S. as a whole. Markets that buck U.S. trends are probably more interesting. Which story would you prefer to write—to read? "Just like the rest of America, folks from XXX head South in the summer for some general sightseeing." Or, "While the rest of America summers in the South, folks from XXX head to the mountains of Pennsylvania."

There are other important factors in understanding the audience for travel journalism. They include:

- The concept of the traveler as a "cosmopolitan"
- Matching travelers and destinations according to "personality"
- The "preferred audiences" of the most popular travel guidebooks and programs
- A pair of key niche audiences: students and seniors
- Using alexa.com to analyze the audience for travel-related websites
- A fast-growing, if nontraditional, niche audience: expatriates

## The Concept of the Traveler as a "Cosmopolitan"

*Cosmopolitan*, according to the nineteenth-century philosopher John Stuart Mill, means belonging to all parts of the world. An increasing number of world travelers, wherever their starting point, perceive themselves as cosmopolitans.

As a practical matter, writes Ulf Hannerz, all this means is "an ability to make one's way into other cultures."[13] But for others it's more a philosophical matter, and it's not universally applauded.

Cosmopolitanism emerged as a way to promote unity among Greek city-states and the cultural unity of the Roman Empire.[14] During the Enlightenment a new form of cosmopolitanism flourished: the shared democratic ideals the Europeans hoped to spread to the rest of the world through colonialism. Modern cosmopolitanism is embodied in the phrase "citizen of the world."

For the traveler, Craig Calhoun writes, this means embracing values that transcend national borders: diversity, globalization, the free exchange of ideas. It means moving from place to place, interaction to interaction, "armed with visa-friendly passports and credit cards." Cosmopolitanism, for these travelers, has "considerable rhetorical advantage"; it is better to be thought of as a "citizen of the world" than to be thought of as a "parochial."[15]

But there are disadvantages, too. Cosmopolitanism tends to be an attitude of elites who embrace—indeed, manage—global capitalism. "Such cosmopolitanism," Calhoun writes, "often joins elites across national borders, while ordinary people live in local communities."

Deep inequalities, even if noted by travelers, are often ignored.

Travel journalists risk neutrality if they embrace cosmopolitanism as an aspect of their identity. While it's fine to draw distinctions, in the reporting, between "elites," "informed citizens," and the "man on the street," it's better not to pick sides.

## Matching Travelers and Destinations According to "Personality"

Marketing researchers have gathered 45 years of personality data about travelers and destinations. The pioneer researcher was the late Stanley Plog, a Harvard-trained psychologist. Plog developed a personality scale[16] to explain how travelers choose destinations.

At one end of the scale is the Venturer, described as one who goes "to more places, more often and participate[s] in more unique experiences than anyone else." At the other end of the scale is the Authentic, described as one who prefers "a life that is more structured, stable and predictable."

In between are lesser degrees of each personality (e.g., Mid-Venturer, Centric-Venturer, etc.).

Plog's research matches traveler personality with destinations. As an example, Venturers match with Hawaii, the Monterey Peninsula, Alaska, Oregon, and Maine in the U.S., and Kenya, Victoria, Antarctica, South Africa, and India outside the U.S. By contrast, Authentics match with Las Vegas, Arizona, Nashville, Honolulu, and South Dakota in the U.S., and Italy, Paris, Australia, Vancouver, and Austria outside the U.S.

Plog's research is important to travel journalists for three reasons:

- First, it's a thoughtful, constructive way to think about the relationship between travelers and destinations. It helps you answer the question: Why do travelers go where they go?
- Second, the travel industry relies on this research as it develops marketing campaigns. It helps you decode their marketing messages.
- Third, it helps you target travelers for destination stories or target destinations for your traveler stories. Don't slant Las Vegas stories to Venturers. Don't slant Authentics to Kenya.

Plog was developing a website based on his research at the time of his death. The finished site is at http://www.besttripchoices.com. One section of the site is Plog's Travel Personality Quiz, 15 questions designed to place you on his scale from Venturer to Authentic. It's quick and fun. More important, it's been scientifically validated over 30 years and 250,000 participants.

Here's a story idea: Ask 15 carefully chosen subjects to take the quiz. Then ask them about where they fell on the scale and what they think of the matching U.S. and foreign destinations.

Here's a sidebar: Take the quiz yourself.

## The "Preferred Traveler" of the Most Popular Travel Guidebooks and Programs

Guidebooks are a "crucial part of the tourist process," writes Deborah Bhattacharyya, "because they mediate the relationship between tourist and destination, as well as the relationship between host and guest."[17]

Lonely Planet is the largest guidebook publisher in the world.[18] As of 2010, it had published about 500 titles in eight languages, and produced television programs, a magazine, mobile applications, and websites.

Rough Guides is a close competitor on many of the same platforms.

Together, they have redefined a huge segment of the audience for travel journalism. Lonely Planet targets what Peter Corrigan calls the "untourist."[19] Rough Guides targets what Maxine Feifer calls the "post-tourist."[20]

These labels, and what they imply, amount to a judgment by Lonely Planet and Rough Guides about the world's "preferred travelers," writes Elfriede Fursich.[21] She analyzed the way these travelers are portrayed on the companies' television programs.

The "untourist" is a backpacker who reflects the spirit of budget traveling. She avoids mass-tourism locations, choosing instead developing or formerly unstable countries far off the beaten track. Travel often is difficult, even arduous. Concerns along the way are typically environmental. Authenticity is key, along with solitude and a focus on basic traveler needs.

"Untourists are critical about touristic practices other than their own," Fursich writes.

The "post-tourist" in Rough Guides doesn't think much of, well, tourism. She visits cities, not small towns or rural areas. She criticizes the political, social, and economic situations in her destinations, Fursich says, stressing human rights, equality of access, and environmental issues. Tourism is a trap, a predicament.

"[T]ourism becomes a questionable pastime [for the untourist] because all problems—social, political, economic and environmental—are related," Fursich says, "and there is no vacation escape possible."

Understanding how Lonely Planet and Rough Guides view their audience, these "preferred travelers," is important to the travel journalist. It's a bit of a chicken-egg issue. Which came first, the audience or the books, programs, and sites that guide them? Likely, they reinforce one another—a virtuous or vicious cycle. Regardless, the combined reach of Lonely Planet and Rough Guides means the travel journalist must understand, and cannot disregard, the "untourist" and the "post-tourist."

## A Pair of Key, Niche Audiences: Students and Seniors

The student audience for travel journalism is large and fast-growing.[22]

Think of the student audience in two segments. One is the school-age segment, roughly ages 12 to 18. They represent a travel market that is "significant in both its scale and economic importance," according to a 2006 study by Michigan State University's Student and Youth Travel Research Institute.[23]

The study found that:

- Some 76.1 percent of parents reported that 12–18-year-olds in their households had taken an independent, overnight group trip.
- Nine of 10 trips were domestic, one in 10 was international. California was the most popular state destination; Washington, D.C. was the most popular city destination.
- More than half the trips were school-related.
- Participation in organized sports was the leading purpose of the trips.

When the researchers extrapolated the data to the total U.S. population of 12–18-year-olds, almost 25 million trips produced $9.8 billion in expenditures.

Then there is the college-age segment.

Estimates vary,[24] but U.S. college students represent as much as 29 percent of international travelers. No other niche audience accounts for as much. Most of these travelers are enrolled in study abroad programs (see Chapter 7 for additional information about study abroad programs). But an increasing number are independent travelers to increasingly remote destinations.

Trends associated with both segments, industry analysts say,[25] include these three:

- Both school-age and college-age students share the sense that "the world keeps getting smaller." For them, this means an ease of "communicating with, doing business with, and traveling to other parts of the world."
- Both school-age and college-age students want to visit destinations common to most travelers, and "they also are more willing to travel to places that have intimidated older generations."
- The risky global political climate doesn't deter either school-age or college-age students. Tour operators and study abroad program directors are adjusting, developing contingencies to allay parents' and schools' concerns for safety.

Foreign travel is the #1 lifestyle choice for higher-income seniors.[26] Only "grandchildren," in second place, comes close.

Just over one-third of households 65 years old and over, earning $75,000 or more, travel outside the U.S. That's more than 1.3 million households. They mostly travel by air and by cruise ship.[27]

They increasingly rely on online services for travel-related activity. Planning leisure travel is the #8 online activity for seniors, according to Nielsen research. Other online activities that may be travel-related include: viewing or printing maps, #2; checking weather, #4; and viewing and posting photos, #5.[28]

Where do seniors travel from? Most often from Florida markets with a lot of well-heeled seniors: West Palm Beach, Fort Myers, and Naples rank #1–#3.[29]

## Using Alexa.com to Analyze the Audience for Travel-related Websites

Alexa.com, a unit of Amazon.com, is a website analytic tool. For the travel journalist's purposes, it's important because it analyzes website audiences by category, and one of the categories is travel.

According to alexa.com's Alexa Traffic Rank, there are 6,363 "top sites" in the travel category. Alexa.com analyzes each site in several ways, including number of unique visitors, page views, bounce rate, and audience demographics.

One reason alexa.com is so useful as a website audience analytical tool is that each category is broken down into several layers of subcategories. As an example, here is the layering I followed in a 2011 search:

Under the category "travel" are 13 subcategories. I clicked on the subcategory "preparation." Under the subcategory "preparation" are eight subcategories.

I clicked on the subcategory "packing." Under the subcategory "packing" are nine websites that, according to Alexa Traffic Rank, are "top sites."

I clicked on the first site on the list, "onebag.com," which promises advice "on the art of traveling light, living on a single carry-on-size bag." Here's what I learned about its audience. Relative to the general Internet population, one.bag.com's audience is:

- Female
- 18–34
- College-educated
- Childless
- Reading the site from work or school

This is the same audience, it turns out, for the nine other "packing" sites.

So, how is this information useful to the travel journalist? It quite precisely tells you who is interested in "packing"—the audience you'd be addressing were you to write a story, or build a competing site, about this anxiety-provoking topic.

This information might be useful to you in a second way. If you were an innovative, entrepreneurial-type travel journalist (see Chapter 6) you might ask, how about the men? Then you might quickly and cheaply test a "packing" site targeting men.

Not so innovative or entrepreneurial? Fine, write an against-the-grain story.

## A Fast-growing, If Nontraditional, Audience: Expatriates

Expatriates are rarely thought of as an audience for travel journalism. In fact, expatriates may be a better *subject* for travel journalism than audience for travel journalism. Regardless, expatriates should be considered.

Let's start with definitions. Expatriates leave their country of origin to live and often work in another country, temporarily. Expatriates fall into one of four groups:[30] skilled technicians and professionals; so-called "lifestyle migrants," often retirees seeking a better climate and lower cost of living; students studying abroad, some for weeks at a time, some for years; and so-called "economic migrants" seeking better jobs at higher wages.

All four groups are difficult to count, but experts agree all four groups are large and fast-growing. The United Nations estimates that more than 200 million people worldwide were living abroad in 2010.[31] That number is

24 percent higher than a decade earlier, 54 percent higher than two decades ago. Globalization is thought to be a key driver of the growth.

All four groups are represented in the U.S. Some are expatriates coming to the U.S.; others are expatriates leaving the U.S.

Expatriates coming to the U.S. are known as "legal temporary residents,"[32] both workers and students. They make up about 3 percent of the country's immigrant population of 38 million people—about 1.1 million.[33]

Expatriates leaving the U.S. number as many as 8 million, according to the U.S. Department of State.

These groups of expatriates—those coming or going—are an audience for expatriate-focused travel journalism, both in the U.S. and abroad. Clearly they are a niche audience, but one that is eager for information about three topics: economics, experience, and offspring. Stories might address the very questions that expatriates often pose:

- Which countries are emerging as economic hotspots for expatriates?
- What is more important to expatriates, personal development or economic gain?
- How well will my children adapt to living abroad?

Similarly, they are a subject of expatriate-focused travel journalism, both in the U.S. and abroad. Think of them as story sources while here or away, who can be consulted to personalize News, Advice, Destination, or Journey stories (see Chapter 3). As a travel opportunity, follow them, observe them, coming here or getting there.

Many U.S. markets have developed a substantial expatriate infrastructure: Find the "Plymouth Rock," the place each group visits first. Investigate the financial services that process remittances to family back home—a common practice of expatriates. What are the legal, medical, political issues?

It should not be difficult to identify expatriates coming to or leaving a U.S. market. Certain employers, typically those with a global footprint, are known to bring expatriates to the U.S. or send them abroad. Consult the Chamber of Commerce. Retiree expatriates are most likely to be located in countries whose characteristics favor expatriates. *International Living* annually ranks countries based on eight criteria.[34] Its top 10 countries for 2010 are Ecuador, Panama, Mexico, France, Italy, Uruguay, Malta, Chile, Spain, and Costa Rica. Start looking there. Colleges and universities keep close track of students coming from abroad and students going abroad. Look for an "office of international education."

# Staying Current about the Audience for Travel Journalism

Audiences aren't static; they change.

Keep up with the size, composition, and behavior of the audience for travel journalism by consulting the sources already mentioned in this chapter: the U.S. Travel Association for the domestic travel industry and the United Nation's World Travel Organization for the global travel industry; the annually updated SRDS *Lifestyle Market Analyst* for demographic and psychographic data by market and lifestyle choice; the periodically updated "Travel Categories" from the Newspaper Association of America; and alexa.com's "top sites" data for the travel category and associated subcategories.

Some travel journalists track audience wants and needs, often at little cost. Some examples include:

- Travel writer and "how-to" travel writing author Louise Purwin Zobel has surveyed broad cross-sections of adults about what kinds of travel articles they like to read in print and online. Two themes emerged: People want "advice, facts, knowledge of how to do it," and they "want to hear about people."[35] The findings reinforce her view that "most of the travel articles we consider good focus on people."[36] A reasonably priced tool for this kind of work is the online survey company zoomerang.com, which offers sampling assistance as well as survey instruments.

- Rolf Potts says understanding audience wants and needs is as simple as scanning the covers of the best-selling travel magazines. "Glossy travel magazines print articles titled 'Sizzling St. Lucia' alongside pristine-looking lagoons," Potts writes, "because—if circulation numbers say anything—travel readers don't want complicated realities: They want a Platonic world unencumbered by crowded beaches and lost luggage."[37]

- Query readers on websites or blogs using the survey widgets. The key is asking engaging questions that are specific, yes or no, timely, edgy, true or false, and fun, says consultant John Haydon.[38] "If they have to spend time trying to understand a question," writes Haydon, "they'll be less likely to answer it."

So, where do we go from here?

You now know some things about the audience for travel journalism. Most importantly, you know to distinguish between the audience for leisure travel and business travel. Whether leisure or business, you know the audience is large and growing. You know about its composition, whether demographic or psychographic. And you know about its behavior—the frequency, mode, and purpose of travel.

Secondarily, you know about the concept of the "cosmopolitan" traveler; how marketers match travelers and destinations according to "personality"; how the "untourist" and the "post-tourist" have emerged as the "preferred audiences" of the world's most popular guidebooks; about such niche audiences as students, seniors, and expatriates; and about alexa.com as the single most important tool for understanding audiences online.

So armed, you must come to terms with a central journalistic question: Who's in charge of decisions about content—the journalist or the audience? It's a question debated by journalists since the 1970s, and it's relevant to the travel journalist today.

Over much of the last century, most journalists were aloof from—even disdainful of—the idea of audience. What does the audience want? Most journalists, focused on providing the audience with what it *needs*, had no interest in the answer.

Leave questions about *wants* to marketing, they said.

But in the late 1970s journalists recognized, perhaps for the first time, that its audience was declining. And, they saw, too, that marketing's voice was stronger than theirs in what to do about it.

So audience, with its questions about wants, gained traction. The answers tended to be polarizing.

One side said that journalists should think of their audience as "citizens."[39] Journalism "provides something unique to the culture: independent, reliable, accurate and comprehensible information that citizens require to be free," write Bill Kovach and Tom Rosenstiel in *Elements of Journalism*. "A journalism that is asked to provide something other than that subverts democratic culture."[40]

The other side said that journalists should think of their audience as consumers. Journalists must become "more reader-driven, customer-driven, looking much more outward and less complacently inward," wrote James K. Batten, then CEO of Knight-Ridder, a newspaper group known for the quality of its journalism. "The balance of power has shifted from editors to readers."[41]

Which side won the argument?

Until a few years ago, the most constructive answer came from Jack Fuller, a journalist, a publisher, and CEO of the Tribune Company: Journalism need not choose between serving the audience as citizen or as consumer. "The question is," Fuller wrote, "how it can best square the two."[42]

But then technology disrupted the business model for newspaper journalism and the bottom fell out of the businesses that fund it. The debate about audience didn't go away, but it was subsumed by questions about survival. The survivors would be online, where indicators like audience reach, engagement, and collaboration dominate.

For the travel journalist today, the audience is best thought of as sitting astride a continuum. On one side are news, enterprise, and investigative coverage of the travel industry. Here the audience is Kovach and Rosenstiel's "citizen." In the middle is advice and service to the traveler as "consumer." And, on the other side, is the "follow me on a journey," where the audience is entertained. For the travel journalist, the key is finding a place on the continuum that squares with his ambition, skills, and ethics. As you decide, there are a few more things to think about audience.

## The Limitations of Audience Research

Research that describes the size, composition and behavior, of an audience is well understood and useful to all journalists, including travel journalists. How many people read your story? What do we know about them demographically, psychographically? How did they use the information you provided?

All useful. But does it tell you what to do next? Probably not.

Descriptive research is like a school report card, or a quarterly earnings report, or the data mined by website analytics: It's about the present moment or the recent past. It does not forecast the future.

As analysts like to say: Data is not the same as insight.

In 1999 journalist Charles Layton wrote a 29-page analysis of newspaper audience research for the *American Journalism Review*.[43] Its central conclusion: Audience research "yields as much uncertainty and confusion as clarity. Much of it is subjective, unscientific and amenable to manipulation."[44] Layton's piece was a harsh judgment, but it highlighted limitations about audience research worth remembering: Quantitative research is subject to a range of biases. They include biases introduced in the wording of comparative questions, in the order of questions asked, and in the characteristics of those who responded and those who did not. The reliability and validity of qualitative research, such as focus groups, is "highly suspect."[45] One big reason:

Samples are small and not representative, yet some may act on what amounts to slim evidence.

# Focusing on the Audience You Have May Impede Finding the Audiences You Need

Successful businesses, whatever their product or service, tend to focus on their best customers—the customers who account for the largest share of revenue, the customers who account for the highest profit margins, the customers who cost the least to serve.

Just makes sense. But there is a pitfall.

Consider the newspaper business. Twenty percent of a newspaper's advertisers account for 80 percent of its advertising revenue. Advertisers in three classified categories—help wanted, automotive, and real estate—cost the least to serve, and account for the highest profit margins. No wonder, then, that when competitors came after the newspaper business, they focused on the classified ads.

Other businesses want your best customers. One way they will compete for them is to "disrupt" your business, usually with a substitute process or technology.

## Test Your Understanding

1. Understanding the audience for travel journalism is important for four reasons. What are they?
2. An important source of data about the composition of the audience for travel journalism is the *Lifestyle Market Analyst* from Standard Rate and Data Service. Describe both "profiles" analyzed by this source.
3. How can travel journalists use the data that underlies Mediamark Research's "Travel Categories" studies for the Newspaper Association of America?
4. What do researchers mean by the concept of the traveler as a "cosmopolitan"?
5. Expatriates are rarely thought of as an audience for travel journalism. Why?
6. Who should be in charge of content decisions in travel journalism—the journalist or the audience? Why?
7. Researchers acknowledge several limitations of audience research. Identify and discuss one of them.

## Practice Your Skills

Assignment #2: Getting to know the local market for travel journalism. Use the *Lifestyle Market Analyst* from Standard Rate and Data Service, available in print or online through a public or academic library, to profile the audience for travel journalism in your market. Answer these questions:

1. In which of the 210 Designated Market Areas (DMAs) is your county (or counties)?
2. What is the total adult population in the DMA?
3. What percentage of the total adult population in the DMA participates in the following lifestyles?
   - Attends cultural and arts events
   - Takes cruise ship vacations
   - Travels in the U.S.
   - Travels outside the U.S.
   - Is a frequent flyer
   - Is interested in gourmet cooking, fine foods, and wine
   - Bicycles frequently
   - Walks for health
   - Hikes and camps
   - Is interested in wildlife and the environment
   - Is interested in photography
4. Do any of the lifestyle interests listed in Question 3, above, index above the U.S. average?
5. What are the demographic characteristics associated with these lifestyle interests? Any in common?

Use this information to profile the audience for travel journalism in your market.

# · 3 ·

# HOW DO YOU REACH THE AUDIENCE FOR TRAVEL JOURNALISM?

Venice, Italy, is among Europe's busiest cruise-ship ports. In 2010 more than 1.6 million visitors arrived by cruise ship, a sixteen-fold increase in two decades. An increasing number of Venetians are angry about this trend. Their concerns: water, noise, and air pollution; potential damage to marble-clad buildings and their underwater foundations; the impact on a delicate and instable ecosystem; crowding.

I know this from the *New York Times*. Rome bureau reporter Elisabetta Povoledo wrote about the controversy in May 2010 in an article titled "Tourist Ships by the Hundreds Rattle Windowpanes and Nerves in Venice." The story played inside the main News section on a Sunday.[1]

Had I been paying attention, I might have known about this issue much earlier from other coverage in the *Times*. Alessandra Stanley first wrote about this phenomenon in 1998.[2] In her story about the arrival in Venice of the world's largest cruise ship at that time, the *Grand Princess*, she wrote that the ship "seemed to eclipse even the cupola of St. Mark's Cathedral." More recently, there was a story about Venice in the Science section of the *New York Times*,[3] and a post on *Green*, a *Times* blog about energy and the environment.[4]

I couldn't, however, have learned anything about this controversy in the newspaper's Travel section. Though Travel covers Venice and cruise ships,[5]

I found no stories about the trends, issues, or controversies raised by coverage elsewhere in the newspaper. Indeed, the most recent feature-length Travel piece about Venice was about seeing the city from the vantage of—a kayak.[6]

The point of this anecdote is not to belittle the newspaper's Travel section for "missing" a story. Rather, the point is to raise central questions facing travel journalists as they figure out how to reach the audience for travel journalism: What is our story? What is our angle? How do we play it, and where? The *Times* "covered" the story—quite fully—but in its News and Science sections, not in Travel.

Why is this a story for news, science, energy and the environment—but not for travel? The question goes to the definition of travel journalism; the identity, purpose, and method of the travel journalist; what the audience for travel journalism wants and needs; and how the travel journalist reaches that audience. Answering the last of these questions—how the travel journalist reaches the audience for travel journalism—is the purpose of this chapter.

The starting point is the question every journalist asks: What's the story? Like almost every other question in journalism, the answer depends on whom you ask. I spent a lot of time studying the kinds of stories travel journalists produce and how they decide what to cover. The answers emerge from a matrix-like decision-making process that considers five choices: story type, topic niches, geography, outlet, and cost.

## The Four Story Types

Travel stories fall within four broad types: News, Service and Advice, Destination, and Journey. Here are descriptions of each type, based on a consensus among practitioners, the dominant "how-to" texts, and the academic literature.

## The News Type

This type embraces news, enterprise, and investigative coverage of travel and tourism, one of the world's largest industries. The News type grows out of the idea that "travel and tourism" is a "beat" like any other—business, sports, health—where news value is judged according to the standard criteria of relevance, usefulness, and interest.[7]

Travel and tourism generate lots of important news and information. It is a global industry, large and growing, employing hundreds of millions. Governments regulate many aspects of the travel industry, including access to and the safety and health of transportation, housing, and food. Governments also tax travel at every level of consumption. And all manner of man-made and natural disasters impact travel.

Here are representative examples of News-type travel stories, drawn from the news archives of the U.S. Travel Association over a recent 15-month period:[8]

- The impact on tourism from disasters such as oil spills, hurricanes, and volcanic ash; from the world economic crisis; from diseases such as swine flu; from terrorism; and from rising gasoline prices
- Travel boycotts inspired by government actions, such as the Arizona immigration law
- Travel industry political clout as seen in passage of the federal Travel Promotion Act, and higher levels of government spending to boost travel, here and abroad
- Scrutiny of corporate and government travel "junkets"
- Airport security, practices of the Transportation Safety Administration, and air traffic control safety
- Jittery valuations of publicly owned travel companies
- Visa waivers

Beyond news coverage are the opportunities for the kind of "accountability" coverage associated with enterprise and investigative reporting. This is reporting "that stems from beat coverage and targets everyone 'who has power and influence over the rest of us'—government bodies, charities, professional sports teams, museums," according to Len Downie, former executive editor of the *Washington Post*.[9] The archives of the nonprofit organization Investigative Reporters and Editors demonstrate that travel and tourism are a rich lode for investigative journalism. Among its recent topics, in order of frequency: tourist safety and health, sex tourism, defrauding of tourists, environmental damage, and government tourism organizations.[10]

## The Service and Advice Type

This type embraces information that helps travelers travel. It grows out of the "service journalism" movement that emerged in the 1970s. This is journalism that emphasizes usefulness, so-called "news you can use." Journalists who practice in this arena often define themselves as "service providers."[11] Travel coverage, for them, is a "service section."[12]

Much of it focuses on the preparation phase of travel, the sort of service and advice that shows up in the "Planning your trip to . . ." sections of travel guidebooks, the "Smart Traveler" departments of travel-related magazines, and the "If You Go" boxes in newspaper Travel sections. Preparedness has become an "obsession" for travelers, writes Jason Wilson.[13] Service and advice-type stories feed on this.

The best of these are the richly sourced, broad surveys associated with travel journalists such as *Conde Nast Traveller*'s Wendy Perrin. Her "Perrin's Reports" are a model for service and advice coverage. As an example, "Perrin's People," published annually, is a global compendium of travel consultants. "Of the more than 10,000 travel agents and tour operators who have tried to break into this list since its inception 11 years ago, only 135 make the grade," Perrin writes. "Getting into this group is, statistically speaking, tougher than getting into Harvard."[14]

Service and advice coverage increasingly focuses on technology, both for its content and for the platform on which it is published. Web applications for desktops and mobile devices increasingly replace human providers of travel service and advice. They range from AutoSlash.com, which searches for low rental car prices, to HopStop.com, which maps local sightseeing trips, to Yapta.com, which tracks airfares and sends email alerts as fares drop. Travel journalists, though they might not define themselves as such, are behind the service and advice on some of these sites. The best example is AirfareWatchdog.com, which promises that its staff will find low airfares that airfare search engines miss.

Yet, some travel journalists disdain service and advice coverage. Jason Wilson, editor of Houghton Mifflin's 10-year-old book series the Best American Travel Writing (BATW), rejects service or advice stories. Readers won't find the phrase "if you go" in any BATW-anthologized story, Wilson writes.[15] One of Wilson's guest editors, Frances Mayes, regards such stories as "mere practicalities."[16]

## The Destination Type

This type embraces the place where you find the story.

Whether nearby or at a distance, this is a travel journalism convention that dates back at least to Herodotus. And therein lies the problem: Every part of the world has been "discovered." There is nothing left to explore. This problem is what was behind Evelyn Waugh's 1946 prediction of the end of travel writing.[17] And it's what led Susan Orlean to conclude in 2007 that travel journalism as exploration is an "outdated notion."[18] Today, the destination story means looking for "places that have changed, or places to visit in a new way," writes Paul Theroux.[19] It means going deeper, farther, not just to find places, but to "root out meanings."[20] This is the work for travel journalists who bring "fresh powers of observation to the [places] they visit," writes William Zinsser, "making us see those places as we have never seen them before."[21]

There is controversy about this type of story. What matters the most: the place, the writer, the people, the story? "Unless there is a strong sense of place," writes Theroux, "there is no travel writing."[22] But what of the writer's experience, asks Zinsser. For him, it's "not what the writer brings to a place, but what a place brings out in a writer."[23] Place is not enough. The destination story requires people. Mark Salzman spent two years exploring Changsha, the capital of China's Hunan Province. "To me, a sense of place is nothing more than a sense of people," Salzman recalls. "Whether a landscape is bleak or beautiful, it doesn't mean anything to me until a person walks into it, and then what interests me is how the person behaves in that place."[24] Place, people or not, is not enough, says Calvin Trillin. The need is for a story.[25] Tim Cahill agrees. "Story is the essence," Cahill writes. "Information is of immense value, but if I can't find a story, I often feel I'm being beaten over the head with an encyclopedia."[26]

Like the rookie pitcher in *Bull Durham*, the travel journalist exploring a destination has to "to learn your clichés." You'll never find a destination story about "the XXX Everyone Knows." Rather, it's the "secret," or the "hidden," or the "what's great about," or the "unexpected." Destination stories won't get by without them. As Pico Iyer notes, the travel journalist "cannot get away with describing the wondrous surfaces . . . better advised to take us into some secret aspect of those places."[27]

Don't be afraid to make judgments, even if it means failing to inspire visits. Some want "down and dirty," according to a *Sydney Morning Herald* blogger, Ben Groundwater. "I don't want to know . . . what the sunlight looked like cutting through the branches of the acacia tree in the Serengeti," Groundwater writes in *The Backpacker* blog. "I want to know what sucked. Give it to me straight."[28]

## The Journey Type

This type embraces getting to the place where you find the story. It derives from one of the oldest plot strategies in literature: the hero's journey. In travel journalism, as in fairy tales, folk stories, and myths, the journey is a cycle: the call to adventure, the journey through the unfamiliar world, the ordeal, the reward, the return.[29]

But with travel journalism, as with all journalism, the journey is nonfiction. What makes journey-type stories compelling? As in fiction, the journey's gone wrong, yet the journalist survives. The best, according to Paul Theroux, relate a journey of discovery that is "frequently risky . . . sometimes grim . . . often pure horror."[30] For Anthony Bourdain, "absurdity is a regular and terrifying feature."[31] Bill Buford wants it to be a journey that "scrambles the writer's brain."[32]

The journey-type story is chronological. "The author's physical progress from A to B makes the factual spine," writes Ian Frazier. "Began here, went there, returned here: came, saw, conquered."[33] Though chronological, the best allow for digression. Indeed, writes Jason Wilson, digression is a part of all great travel writing. In many ways, the digressions are the story."[34]

The point of view is first person: I go, I see, I tell you what I saw.[35] This is armchair travel for the audience, joining the travel journalist vicariously. The audience wants to experience, to feel "the emotional and intellectual weight of the writer's observations," writes Susan Orlean.[36] The chief requirement is the journalist's "opening the imagination to other lives and languages"[37]

The risk with this story type, of course, is a journey not worth sharing.

## How Travel Journalists Value, Apply the Dominant Story Types

Travel journalists, whether editors or reporters, largely agree that these are the dominant story types. I know this from surveys I conducted in 2011 with 139 members of the Society of American Travel Writers, the leading organization of travel journalists in the U.S.[38]

I wanted to know how travel journalists rank the relative importance to readers of the different types of travel stories, and how frequently travel journalists cover the different types of travel stories. Here is what travel editors had to say about relative importance to readers of the different types of travel stories:

- Twenty-nine percent who responded said "news of the travel and tourism industry" is important. Another 29 percent said it is somewhat important, 6 percent said it is unimportant, 24 percent said somewhat unimportant, and 12 percent said neither important nor unimportant.

- Fifty-three percent who responded said "advice, reviews, to help readers plan travel" is important, 24 percent said somewhat important, 12 percent said unimportant, 6 percent said somewhat unimportant, and 6 percent said neither important nor unimportant.
- Fifty-three percent who responded said "guides to U.S. destinations" are important, none said somewhat important, 24 percent said unimportant, 6 percent said somewhat unimportant, and 18 percent said neither important nor unimportant.
- Forty-seven percent who responded said "guides to non-U.S. destinations" are important, 12 percent said somewhat important, 18 percent said unimportant, 6 percent said somewhat unimportant, and 18 percent said neither important nor unimportant.
- Twenty-four percent who responded said "accounts of personal journeys" are important, 24 percent said somewhat important, 18 percent said unimportant, 18 percent said somewhat unimportant, and 18 percent said neither important nor unimportant.

What's to be drawn from this data?

Nearly half to three-quarters of travel editors find each of the story types to be important or somewhat important. "Advice, reviews, to help readers plan travel" ranks highest in importance. This reinforces Jason Wilson's observation that we have reached "a gilded, Rococo age" of this type of service-oriented travel journalism.[39] "Accounts of personal journeys" ranks lowest, reflecting perhaps its literary ties, which many journalists find suspect, or its entertainment purpose, which many journalists disdain.

It should come as no surprise that the freelance writers who pitch their work to these editors seem to agree about the relative importance of these story types. I asked them how often they cover the same types of travel and tourism stories. Here's what they said:

- Twenty-five percent who responded said they frequently cover "news of the travel and tourism industry," 37 percent said infrequently, 38 percent said never.
- Sixty percent who responded said they frequently cover "advice, reviews, to help readers plan travel," 30 percent said infrequently, 10 percent said never.
- Fifty-nine percent who responded said they frequently cover "guides to U.S. destinations," 27 percent said infrequently, 14 percent said never.

- Fifty-three percent who responded said they frequently cover "guides to non-U.S. destinations," 28 percent said infrequently, 18 percent said never.
- Sixty-five percent who responded said they frequently cover "accounts of your travel journeys," 27 percent said infrequently, 8 percent said never.

What conclusions can be drawn from these answers?

There's a fairly strong alignment between the types of stories editors think readers value and the types of stories freelancers produce. Makes sense, given the symbiotic relationship between travel editors and freelancers. The editor buys from the freelancer; the freelancer's income depends on an editor's buying decision. Only the frequency with which freelancers cover "accounts of your travel journeys" is nonaligned: 65 percent of freelancers who responded cover this story type frequently, yet only 24 percent of editors who responded think this story type is important to readers.

## The Twelve Topic Niches . . . and Counting

The four story types are the starting point in deciding how to reach the audience for travel journalism. The next step is topics. Some topics are travel journalism antes: You cover them in order to be "in the game." They are the fodder of generalized guidebooks: when to travel, how to get there, getting around, where to stay, what to eat, the most important sights and attractions, and so forth. Topic niches, by contrast, are tuned to the audience's specialized interests. Wikipedia hosts entries on 45 tourism niches from accessible tourism to wine tourism. Niches change over time. Sometimes the audience wants adventure. Then the audience craves food and wine. Maybe next, the audience wants volunteer opportunities. Then cruising. Audiences go back and forth in an effort to fulfill what they wants from travel: traditions and authenticity, intimacy and connection, sociocultural knowledge, and—increasingly—the chance to help (at least not to harm) the places that give us pleasure.[40]

Twelve topic niches vary in importance to readers, according to the travel editors who responded to my survey. In order of importance, they are:

- Cruising
- Adventure
- Food and wine
- Heritage

- Eco
- Medical
- Sex
- Volunteer
- Poverty
- Gay and lesbian
- Genealogical

Here's an introduction to each niche topic. Generally speaking, they can be applied to all four story types. For example: News of the cruise industry, Service and Advice about off-shore medical providers, a Destination that is a heritage site, a Journey along a genealogical pathway. But some niches work better with one or more story type, and work less well with others.

Each one can be reported to a high journalistic standard.

## Cruising

Some 47 percent of travel editors who responded to my survey said "cruising" is an extremely important niche topic to readers. Another 12 percent said it is somewhat important. Here's information about the audience for this niche, one angle to pursue,[41] an example of best-practice coverage, and where to learn more.

- Audience: Some 18.3 million U.S. households regularly take cruise vacations.[42] Cruise vacationers, compared with all U.S. households, are older, more often married, and higher income.[43] Cruise vacationers are likely to be "centric venturers" who see cruising as affordable, easy, and comfortable.[44] Most are U.S. and Canadian, but an increasing number are Europeans. Key to niche: More than 90 percent of U.S. adults have never taken a cruise.[45]
- One angle to pursue: Strong opportunity for news, enterprise, and investigative coverage. The cruise industry is large, growing, and controversial. It was page-one news for days in 2010 when the *Carnival Splendor* lost all power, stranding 3,000 passengers at sea. Among newsworthy angles: passenger health and safety, air and water pollution, and impact on ports.
- An example of best-practice coverage: Kristen Bellstrom's "Cruise Ships on Steroids" (*SmartMoney*, November 6, 2009) won the Gold

award for an article on marine travel in the 2010 Lowell Thomas Awards. This is a richly sourced, brightly written piece on cruise-ship industry expansion during economic recession. The judges said: "This piece gives us much more than oohs and ahhs about over-the-top cruise ship extravagances. It examines the quest to outdo what's been done before in shipbuilding and reveals what this arms race is doing to the cruise line industry and passengers."[46]

- To learn more: *Cruise Week News* covers the industry. *Cruise Critic* reviews ships, itineraries, and ports of call. *World Cruise Industry Review* is the industry's North American glossy trade publication. "Marine" coverage is a category in the Lowell Thomas Awards. Among top winners in the last 10 years were: the *Chicago Tribune*, the *San Francisco Chronicle*, the *Washington Post*, and *National Geographic Traveler*.

# Adventure

Some 41 percent of travel editors who responded to my survey said "adventure" is an extremely important niche topic to readers. Another 35 percent said it is somewhat important. Here's information about the audience for this niche, one angle to pursue, an example of best-practice coverage, and where to learn more.

- Audience: Some 26 percent of respondents in a 2010 study said they participated in adventure travel.[47] Virtually all engaged in "soft" adventure activities such as archeological expeditions, bird watching, hiking, and safaris. Only a handful engaged in "hard" adventure activities such as caving, trekking, and climbing. Adventurers are equally likely to be single or married, male or female. Compared with other travelers, they are better educated and higher income. They place a higher importance on exploring new places, time in nature, engaging with local cultures, pushing physical limits. More than half of their adventures are in the continents where they reside. Top three destinations among developed countries are Switzerland, Sweden, and New Zealand.
- One angle to pursue: Adventure travelers, compared with all travelers, spend significant amounts of money on equipment and apparel before they travel.[48] Their "gear budget" ranges from 43 percent of travel costs for "soft" adventures to 87 percent for "hard" adventures.

So, interview adventure-travel outfitters in your market about trends in equipment and apparel. Ask a trio of experienced adventure travelers in your market to test and review the "latest" equipment and apparel. But also ask them to tell you about the oldest, most reliable piece of gear in their kits.

- An example of best-practice coverage: Patrick Symmes' "Hugo's World" (*Outside*, April, 2009) won Gold for an article on adventure travel in the 2010 Lowell Thomas Awards. The judges said: "Patrick Symmes' piece on traveling the wilds and the cities of Venezuela takes the reader across this country's political and cultural landscape, as well as its literal one. The writer's mix of practical advice, curious details and historical perspective makes this an outstanding story."[49]
- To learn more: Adventure travelers rely on *National Geographic* and *Conde Nast Traveller*.[50] The Adventure Travel Trade Association (ATTA) underwrites significant amounts of research, and its monthly e-newsletter, *Adventure Travel News*, is free. The International Institute for Tourism Studies at George Washington University's School of Business and Xola Consulting are ATTA's academic partners, responsible for much of the research. "Adventure" coverage is a category in the Lowell Thomas Awards. Among top winners in the last 10 years were: *National Geographic Traveler*, *Outside*, *National Geographic Adventure*, *Adventure Cyclist*, the *Boston Globe*, and the *Los Angeles Times*.

## Food and Wine

Some 29 percent of travel editors who responded to my survey said "food and wine" is an extremely important niche topic to readers. Another 35 percent said it is somewhat important. Here's information about the audience for this niche, one angle to pursue, an example of best-practice coverage, and where to learn more.

- Audience: Some 24 million U.S. households say gourmet cooking and fine food represent the "good life."[51] A couple of million fewer say the same about wines.[52] Food and wine aficionados, compared with all U.S. households, are middle-aged married couples with two higher-than-average incomes. Their other lifestyle interests are attending cultural events, foreign travel, and investing in stocks, bonds, and real estate.

- One angle to pursue: Have some fun with one or more "book of lists" approaches to food and wine in your market. Try a "best of . . . ," "hidden . . . ," "secrets . . ." approach to food and wine in your market. Who are the best, at-home, regional cooks? Who have the best wine collections? Try a scavenger hunt for the "best of . . . ," "cheapest . . . ," "most obscure . . ." foods and wine in your market. Where can you buy locally smoked pig parts? Who sells the cheapest bottle of DOCG Chianti? Or, is there a locavore movement in your market? Take a hard-nosed look. Or, look at health inspections. Or, what food is grown in a 50-mile radius of your market.

- An example of best-practice coverage: Steven Rinella's "Me, Myself, and Ribeye" (*Outside*, April 2009) won Bronze for an article about Argentina beef in the 2010 Lowell Thomas Awards. The judges said: "Steven Rinella uses sharp and self-deprecating humor to describe the people who farm, butcher and grill cattle in Argentina's meat-heavy food chain. He equates the obsession and difficulty in searching for the best steak in Argentina to 'trying to pinpoint the whereabouts of Osama Bin Laden.'"[53]

- To learn more: Food and wine travel writing is specialized enough that it has its own association: the International Food Wine and Travel Writers Association. A $150-per-year membership provides access to its highly useful monthly magazine, *Press Pass*. Careersinfood.com has a simple-to-use, robust search engine on 29 food and wine topics. The category "Food and Beverage Trade Associations," for example, lists more than 100 organization websites, from Agriculture Council of America to Wine Institute. Searches can be filtered by keyword and zip code.

# Heritage

Some 18 percent of travel editors who responded to my survey said "heritage" is an extremely important niche topic to readers. Another 35 percent said it is somewhat important. Here's information about the audience for this niche, one angle to pursue, an example of best-practice coverage, and where to learn more.

- Audience: Some 7.6 million U.S. households regularly engage with "our nation's heritage."[54] They are a "geeky" lot, also interested in science and technology, science fiction, and coin and stamp collecting.

Compared with all U.S. households, they are middle-aged married couples with two higher-than-average incomes.

- One angle to pursue: This niche tends to be anniversary driven. Is an anniversary approaching that's relevant to heritage in your market? As an example, 2011–2015 is the sesquicentennial of the beginning and end of the U.S. Civil War. All manner of destination marketing organizations are attempting to attract tourists to their Civil War venues. The Civil War Trust has assembled a "to do" list of 150 Civil War activities—some are historical sites to visit, others are activities such as holding a mini ball or watching the film *Gone with the Wind*. What from this "to do" list can you find in your market? If nothing, assemble your own "to do" list, but only real things. "There's nothing more powerful than the authenticity of the real thing," writes Gary Adelman. "We stand awestruck at these places, as we try to learn what happened there, and we are moved in some way—whether saddened, angered, confused, uplifted, passionate, or enriched."[55] Is an anniversary approaching that's more relevant to heritage in your market than the Civil War? If so, apply this approach to it.
- An example of best-practice coverage: Abigail Tucker's "Road Music" (*Smithsonian*, September 2011). Every issue of *Smithsonian* includes one or more stories that fit the topic niche, heritage tourism. Tucker's is typical: rich, deep, nuanced reporting about the "crooked tunes" that enliven mountain life in southern Virginia.
- To learn more: Many markets have local history museums. Most public libraries employ reference librarians who specialize in local history. The History Department at most local universities usually has one or more members of the faculty who specialize in local history.

## Eco

Some 12 percent of travel editors who responded to my survey said "eco" is an extremely important niche topic to readers. Another 41 percent said it is somewhat important. Here's information about the audience for this niche, one angle to pursue, an example of best-practice coverage, and where to learn more.

- Audience: Eco-tourism is an example of a "soft" activity of adventure travel,[56] so the audience for this niche is some subset of the 26 percent of a study's respondents who said they participated in adventure travel. Indeed, the audience for this niche may be much higher: 79 percent of U.S. adults consider themselves to be environmentally conscious and increasingly aware of terms such as *carbon footprint* and *global warming*.[57] And some 17.3 million U.S. households say wildlife and environmental activities are ways they regularly engage the "Great Outdoors."

- One angle to pursue: Consider an apparent paradox: While environmental responsibility is a prime factor influencing selection of travel providers, according to the U.S. Travel Association, travelers are unwilling to pay extra to support environment-friendly travel.[58] Interview local travelers about the environment-friendly practices they observe. Test whether they are spending more to travel according to environment-friendly practices. Do they, for example, offset their carbon footprint? Use one of the travel-oriented carbon footprint offset calculators[59] to show them the cost of offsetting a recent trip. Are they willing to pay? How many offset the carbon footprint of their last flight? Many airlines offer this option when booking.

- An example of best-practice coverage: Jeff Greenwald's "Bedeviled Island" (*Afar*, 2009) won the Gold for an article on environmental tourism in the 2010 Lowell Thomas Awards. The judges said: "Jeff Greenwald does a masterful job of making the readers care about Tasmania Island by making us care about a strong central character (Geoff King) and his decision to turn his cattle ranch into a wildlife preserve. The move ostracized King from his neighbors and his own family, as the fate of the area remains a divisive issue as proponents on both sides struggle to find the correct balance between environmentalism and tourism."[60]

- To learn more: The International Ecotourism Society[61] is a 20-year-old nonprofit "dedicated to ecotourism as a tool for conservation and sustainable development." Its "experts bureau" is an excellent resource for travel journalists. Another source is the Adventure Travel Trade Association mentioned under the Adventure niche above. "Environmental Tourism" coverage is a category in the Lowell Thomas Awards. Among top winners in the last 10 years were: *National Geographic Traveler*, *National Geographic Adventure*, *Outside*, *Via*, and *Islands*.

# Medical

Some 6 percent of travel editors who responded to my survey said "medical" is an extremely important niche topic to readers. Another 24 percent said it is somewhat important. Here's information about the audience for this niche, one angle to pursue, an example of best-practice coverage, and where to learn more.

- Audience: Estimates vary sharply about the audience for medical tourism. Deloitte Consulting, in a widely quoted study, estimated that 560,000 U.S. residents went abroad for medical and surgical care in 2008, and predicted that number will rise to 1.6 million by 2012.[62] One researcher placed the global number of medical tourists at 5 million.[63] The consulting firm McKinsey, by contrast, placed the number at 60,000 to 85,000, acknowledging that its estimate is "far smaller than others have reported."[64] Whatever the disputes about numbers, all agree that medical tourism is an important, fast-growing aspect of global health care. And it's attracted considerable interest among travel journalists.
- One angle to pursue: Most medical tourism stories are about U.S. residents seeking treatment abroad. Locate several from your market and interview them about their experiences. How to find? Travel agents might be a help, but a better source are the firms that facilitate medical travel. Forty are listed as members of the Medical Travel Association—a good starting point. Or, flip the angle of view, and report about non-U.S. residents traveling for care in the U.S. The Mayo Clinic, as an example, has attracted non-U.S. patients for more than 100 years.[65] How about providers in your market?
- An example of best-practice coverage: Adam H. Graham's "Worst Medical Tourism Disasters" for *Travel + Leisure* in November 2009 offers more than the headline implies. The "botched job" surgeries it documents justify the article's advice: caveat emptor. But the "real deal" advice it offers is characteristic of the best travel-related service journalism.
- To learn more: The Medical Tourism Association is the industry's trade association.[66] It describes itself as a nonprofit, membership-based association of providers. It archives coverage of the medical tourism market and publishes an e-newsletter and two magazines. Academic researchers present their work at conferences hosted by the Center for Medical Tourism Research, housed in the school of business at the University of the Incarnate Word in San Antonio, Texas.

# Sex

Some 6 percent of travel editors who responded to my survey said "sex" is an extremely important niche topic to readers. Another 6 percent said it is somewhat important. Here's information about the audience for this niche, one angle to pursue, an example of best-practice coverage, and where to learn more.

- Audience: There are two audiences for coverage of sex tourism. One is the patron audience.[67] The other audience is everyone who is appalled by their behavior. Of the websites that target the patron audience, worldsexguide.com is the best known. But with a couple of exceptions, mainstream guidebooks are on the appalled side of the coverage, with headlines such as "A dark side of . . ." and leads such as "What is even more tragic . . . ." Most entries about the sex trade focus on avoidance, cautions about disease, and whom to notify when travelers suspect children are involved. One exception to this view is guides to Las Vegas, the other is guides to Bangkok. Sex, it would seem, is such a mainstream pursuit of travelers to those cities that "how-to" guidance seems appropriate. Fodor's guide to Las Vegas, for example, devotes 20 pages to a chapter titled "Sex, Lies and Las Vegas."[68] More often, coverage of sex tourism is news, enterprise, and investigative. Sex tourism, for example, ranked second (behind tourist safety and health) in an archive of investigative coverage of "tourism" in the story archive of Investigative Reporters and Editors.[69]
- One angle to pursue: Ride along for one night with a police vice unit. (Most police departments encourage this in programs typically called "Citizen Ride-Along." Contact your department's community relations office.) The opportunity is for a travel narrative interrupted by analysis, what writing coach Roy Peter Clark calls "broken-line narrative."[70] See Chapter 4.
- An example of best-practice coverage: Sean Flynn's "Where They Love Americans . . . for a Living," published in *GQ* in March 2007, was selected for *The Best American Travel Writing 2006*. It's about sex tourism in Costa Rica. Though stylistically a travel narrative, Flynn documents Costa Rica's sex industry in the context of the global sex trade.
- To learn more: In the U.S., consult the Justice Department's Child Exploitation and Obscenity Section.[71] UNESCO's Trafficking Statistics Project found that a dozen organizations across the globe

gather data on trafficking.[72] The most cited source is the U.S. State Department's annual report.[73] A first-rate primer with a strong list of related articles, government agency reports, and NGO contacts was broadcast in 2006 by *Frontline*.[74]

# Volunteer

None of the travel editors who responded to my survey said "volunteer" is an extremely important niche topic to readers, but 35 percent said it is somewhat important. Here's information about the audience for this niche, one angle to pursue, an example of best-practice coverage, and where to learn more.

- Audience: "More and more people in all stages of life are thinking of becoming 'voluntourists,'" says UC San Diego researcher Bob Benson. A 2008 study by his Center for Global Volunteer Service concluded that 40 percent of Americans are willing to spend several weeks on vacations that involve volunteer service.[75] Among their volunteering priorities: connecting with people rather than organizations, and helping school children and the poor. A 2008 study for msnbc.com and *Conde Nast Traveller* found that 20 percent of respondents had taken at least one volunteer vacation and 62 percent said they were very likely or somewhat likely to do so.[76] Voluntourism also is an example of a "soft" activity of adventure travel,[77] so the audience for this niche may be some subset of the 26 percent of a study's respondents who said they participated in adventure travel.
- One angle to pursue: Who comes to your market on vacation to volunteer? A likely cohort are the groups of high school and college students organized by Habitat for Humanity. Habitat has spring break and summer programs in 200 communities in 36 states. The students spend a week building a home for a poor family. Use this as a lens to view voluntourism from two standpoints: the donors and the recipients. Questions to ask: How meaningful to the donor, the recipient? Is it better to write a check for the cost of the travel? What is the value of the human connection? Is it lasting?
- An example of best-practice coverage: J. B. MacKinnon's "The Dark Side of Volunteer Tourism," first published by the Canadian magazine *Explore* (2009) and later in the *Utne Reader* (2011). MacKinnon tests

the bounds of volunteer tourism on a trip to Malawi. Despite the title, this is a balanced account. MacKinnon weaves questions about his own behavior with that of others similarly engaged.

- To learn more: An excellent primer is *Frommers'* "How to Plan a Volunteer Vacation in Six Steps," by Melina Quintero.[78] VolunTourism.org is an excellent general site, organized according to your purpose: travelers, travel planners, educators.

# Poverty

None of the travel editors who responded to my survey said "poverty" is an extremely important niche topic to readers, but 12 percent said it is somewhat important. Here's information about the audience for this niche, one angle to pursue, an example of best-practice coverage, and where to learn more.

- Audience: Poverty tourism—and it goes by many names[79]—is the most controversial travel journalism topic niche. The "crux of the debate is whether it's OK to pay to look at poor people," writes Saundra Schimmelpfennig in the introduction to a compilation of commentary on the topic.[80] One side of the argument proposes that traveling among the world's poor encourages compassion and donations. The other side argues it is exploitative and unhelpful. Some of the world's poorest countries encourage tourism. Poverty tourism is a topic of academic research[81] and a recurring—if controversial—topic among travel journalists. It's one of the "slippery moral dimensions of travel journalism," writes Anthony Bourdain. Why? For Bourdain, it's a matter of the travel journalist "becoming complicit . . . in something very, very bad."[82]
- One angle to pursue: Use U.S. Census data to identify the poorest and wealthiest tracts in your market. Visit them at different times of the day, different days of the week. How are they different? How are they the same? How do they intersect? As a sidebar, how did the experience affect your opinion about this topic niche? Do you agree with the critics, or the advocates?
- An example of best-practice coverage: John Lancaster's trend piece "New Stop Squalor," first published in *Smithsonian*, was selected for *The Best Travel Writing in America 2008*. Lancaster explores the issue

of poverty tourism through the lens of one of the most controversial tours: the Dharavi squatter settlement in Mumbai, India.

- To learn more: Academic sources are probably the best opportunity to learn more. One of the best is C. Michael Hall's *Pro-Poor Tourism: Who Benefits?* Cases and reviews assess the effectiveness of pro-poor tourism as a development strategy and confront the issue of who benefits from tourism's potential role in poverty reduction.[83]

## Gay and Lesbian

None of the travel editors who responded to my survey said "gay and lesbian" is an extremely important niche topic to readers, but 12 percent said it is somewhat important. Here's information about the audience for this niche, one angle to pursue, an example of best-practice coverage, and where to learn more.

- Audience: Estimating the size of the audience for gay and lesbian travel is as difficult as estimating the size of the gay and lesbian population.[84] That said, there is a substantial travel infrastructure that targets gays and lesbians. The International Gay and Lesbian Travel Association, founded in 1983, has 2,200 member agents in 83 countries.[85] Virtually every mainstream destination guide contains advice targeting gay and lesbian travelers. The market-research and consulting firm Community Marketing Inc. counsels the tourism and hospitality industry about reaching the audience for gay and lesbian travel. Its "Annual Gay & Lesbian Tourism Report," now in its 15th year, is closely followed.[86]
- One angle to pursue: How is your market similar or different to markets that target gay and lesbian travelers? Use the "Annual Gay & Lesbian Travel Report"[87] as a roadmap. Where does your market rank as a destination? What brands in your market target gay and lesbian travelers? Is there a "pride" event in your market—a strong driver of gay and lesbian travel?
- An example of best-practice coverage: The first-person trend story in the *New York Times*, "After Gay Marriage, the Check-In Dance." Eric Marcus uses his experience with partner Barney as a lens through which he explores the "awkward moments on those occasions when you'd hope to be treated just like every other married couple."[88]
- To learn more: Demographer Gary J. Gates of the Williams Institute at UCLA School of Law is a leading authority on the demographics

of the LGBT community. Data from the U.S. Census is increasingly useful, especially with respect to the number and location of same-sex couples. The International Gay and Lesbian Travel Association represents travel agents that target gay and lesbian travelers. Community Marketing Inc. counsels hospitality and tourism businesses that target gay and lesbian travelers.

# Genealogical

None of the travel editors who responded to my survey said "genealogical" is an extremely important topic niche to readers, and only 6 percent said it is somewhat important. Here's information about the audience for this niche, one angle to pursue, an example of best-practice coverage, and where to learn more.

- Audience: Genealogical tourism is one of the fastest growing segments in leisure travel, researchers say, "because it represents a conscious shift away from relaxation and into the realm of personal enrichment and fulfillment."[89] Baby boomers are the primary audience driving this niche.[90] "Aging plays an important role in defining a person's choice of tourism," writes one of the researchers, Grace Yan, and "genealogical travel is contemporary society's way of attaining a more coherent and continuous, albeit imagined, view of themselves in connection with the past."[91] Genealogical travelers gravitate toward countries such as Ireland that have experienced mass emigration over long periods of time. More than 116,000 "genealogical visitors" ventured to Ireland in 2000, the peak year, according to the Irish Tourist Board.[92] Technological advances have reduced genealogical tourism in Ireland: There were just 45,000 "genealogical visitors" in 2004, the year the Irish Tourist Board stopped counting.[93] Why visit, it seems, if you can trace your ancestry online? Genealogical research services are common online. Ancestry.com is the largest and best known. Some companies facilitate genealogical travel. One of them is ancestralstory.com.[94]
- One angle to pursue: How about a local version of *African American Lives*—on a shoestring? Harvard professor Henry Louis Gates, Jr. produced this television program that features cutting-edge genetic analysis that locates participants' ancestors in Africa, Europe, and America. Later, Gates and his partners formed a company that tests DNA via the same method used on the award-winning television

program. For as little as $189, one can test the origin of either a paternal or maternal line. So, ask a diverse group of prominent locals to be tested—and pay for the test. Then report on the findings. Who knows: Maybe one of them, like President Obama, whose roots were in Moneygall, Ireland, will be surprised enough by the results to travel there—an even better story. Another approach is to use ancestry.com to replicate NBC's approach on the television program *Who Do You Think You Are?*[95]

- An example of best-practice coverage: The AP's Jennifer Dobner produced a strong example of service-and-advice type travel journalism in her piece about genealogy tourists who flock to the Family History Library in Salt Lake City each year. "There's nowhere else," a genealogical tourist from New Zealand told her. "Just nowhere else."[96]

- To learn more: Ancestry.com is the world's largest for-profit genealogical research firm. On the nonprofit, free-to-use side, consult the Family History Centers run by the Church of Jesus Christ of Latter-day Saints. Its databases contain more than 6 billion ancestral records. Also consult the reference librarians in the genealogy sections of local public libraries.

## Other Niche Topics Worth Considering

- Religious pilgrimages—An archetype of the Journey is the travel journalist as pilgrim, or as chronicler of the pilgrim's experience. It's the search for moral or spiritual significance that underlies the pilgrimage—the quest. Pilgrims and their pilgrimages can be found almost anywhere—certainly in all of the world's religious centers. For an example of best-practice coverage, see Robert F. Worth's "Modern-Day Pilgrims Find Interfaith Bond in Ancient Monastery" (*New York Times*, January 19, 2010).

- Dark tourism—This is travel to places associated with death, suffering, or grief. Uncommonly morbid? Not really. Battlefields are the most common dark tourism venues, but there are many others. Among examples cited by *The Guardian*: The world's five most popular graveyards, a former POW camp, the radioactive wasteland of Chernobyl.

- Issues-oriented tourism—Travel journalist Larry Bleiberg is the creator of civilrightstravel.com, a destination guide to Civil Rights–era venues, primarily in the Southeast. This is an example of an issues-oriented travel niche. Any issue would be fair game. An excellent

opportunity for crowd sourcing: ask readers to suggest venues associated with the issue, as Bleiberg does for his site.

- Garden tourism—Nearly 70 million U.S. households say flower and
vegetable gardening are central to their home lives.[97] No wonder,
then, there are some 3,000 garden-related events and festivals in the
U.S., according to Richard Benfield, author of a forthcoming book on
garden tourism. Benfield estimates that garden tourism is one of the
country's fastest growing tourism niches.[98]

- Maternity tourism—Perhaps the newest of the niches, maternity tourism is the practice of non-U.S. pregnant women visiting the U.S. to
give birth, thus conferring U.S. citizenship on their newborn. The
practice is wrapped up in the broader debate about U.S. immigration
policy. For an example of best-practice coverage, see Jennifer Medina's
"Arriving as Pregnant Tourists, Leaving with American Babies" (*New
York Times*, March 29, 2011).

# Reaching the Audience for Travel Journalism

This chapter explores how the travel journalist reaches the audience for travel
journalism. At the beginning of the chapter, we said the starting point of an
answer is the question every journalist asks: What's the story? Like almost
every other question in journalism, the answer depends on whom you ask.
Over the last several pages we've developed a good sense about the *types* of
stories travel journalists produce as well as the *niche topics* they tend to explore.
There are three more elements to consider.
    They are:

- Providing what the audience needs and wants—when, where, and
how it wants it
- Adopting an approach to publishing: anticipating the "great mashup"
- Completing the decision-making matrix: geography, outlet, and cost.

## Providing What the Audience Needs and Wants—When, Where and How It Wants It

Reaching the audience for travel journalism means providing the audience
with what it needs and wants—when, where, and how it wants it. Doing this is

vastly more complicated today. The "what it wants and needs" hasn't changed all that much. It's the "when, where, and how" that's different.

Of course, there are travelers who follow Arthur Frommer's advice to "get lost in the world." Travel journalism, regardless of story type or topic niche, is largely irrelevant to them. But this audience is so small it is to be dismissed. Remember, even "untourists" read the Lonely Planet guides, "post-tourists" the Rough Guides.

The long-practiced four story types still cover the "wants and needs," however the travel journalist elects to tell the story, whether written as news reports, narratives, FAQs, lists, and tip sheets; or expressed graphically as charts, graphs, and maps; or photographically as stills, sound slides, and videos.

The "when" is increasingly real time, regardless of time zone. Digital media encourages impatience. If they are in sight of a mobile transmission tower, travelers want the download now. Not only does the audience want real-time access to information, it wants the information to be up to date—also in real time. Static sources are anachronistic. We understand this intuitively about printed material. Printed guidebooks, as Tom Brosnahan has noted, are out of date *before* the guidebook is published.[99] Articles printed in newspapers and magazines are out of date soon after. But online sources are no better—though they *seem* better—unless they, too, are up to date.

"Where" is everywhere. Service and advice in preparation for travel is wanted at home, school, and office, according to audience research for travel-related websites. Depending on the type of travel, travelers select and book travel from school and office as often as they do from home. Travel-related news, enterprise, and investigative coverage is wanted wherever travelers consume news generally. Destination-type stories are wanted in locations where travelers prepare as well as where they travel. Recall the number of travelers you've observed on the streets of any city or town, thumbing their way through a guidebook. Only Journey-type stories are wanted in one place: wherever the armchair traveler prefers to read.

"How" is fast-changing. The audience for travel journalism in print is turning to online sources, particularly for pre-travel service and advice and destination guides. Sales of printed travel guidebooks are in steep decline,[100] as is circulation of the newspapers with the most honored Travel sections. One of the most honored travel magazines, *National Geographic Adventurer*, went out of business for a lack of advertising. Travel on television is strengthening, with the dominant provider, the Travel Channel, available in 96 million U.S. households.[101] Some 39 percent of all travel transactions are now made online[102] from a desktop, a laptop, or a mobile device. Multicategory sites—led by Expedia—account for half of the 10 most popular travel sites, while map sites account for four of the 10 most popular travel sites.[103] Mobile devices are globally ubiquitous. Nearly one-third of mobile devices are so-called "smartphones," that is, devices with app-based, web-enabled operating systems.[104] Mobile is the leading platform for travel-related innovation.[105] "Daily deals" are the leading innovation.[106] On a recent stopover in Hong Kong, travel-related vendors had texted 13 offers to my mobile device before I had cleared the jetway.

## Adopting an Approach to Publishing: Anticipating the "Great Mashup"

We'll talk more about approaches to publishing in Chapter 5. For now, it's only important to understand that traditional approaches to publishing in print, as noted briefly above, are in decline and giving way to publishing online. Until the 1990s, it was publish either in print or online; by the 2000s, it was publish in both. Today, such thinking about choices seems quaint.

The critic David Carr says we're near the point where there are no verticals—no television, print, web, and radio. "What if they were all just one big blob of media?" Carr asked. "Well, if you are staring at an iPad or some other tablet, that future seems to have already shown up."[107]

It's certainly shown up in travel journalism. The Lonely Planet "team," for example, writes, photographs, and edits travel information that is "published" as books, magazines, newsletters, and programs in print, broadcast, and online for desktops, laptops, tablets and smartphones. There are blogs and forums. Travel news is aggregated. There's an online shop that sells and fulfills. It's one-to-one or many-to-many.

So, as you think about reaching the audience for travel journalism when, where, and how it wants it—think, as Carr sees it, of a "great mashup."[108]

## Completing the Decision-making Matrix: Geography, Outlet, and Cost

We said earlier that travel journalists decide what to produce using a matrix-like decision-making process that considers five choices: story type, topic niches, geography, outlet, and cost. We've talked about the first two—story type and topic niches. What about the last three?

Travel journalism is typically grounded in a place. Service and Advice stories are about preparing to visit a place. Destinations are places. Journeys are about getting to a place. So the third question the travel journalist must answer is about place, or geography. Travel or stay at home? If travel, where? Be a "travel anywhere" generalist? Specialize in one place? How to decide? Fortunately, the decision is not either/or. Start at home, perhaps, with what you know—but reaching for an audience that doesn't. Venture forth to a place where you share a language, customs, infrastructure. Then, perhaps, venture more widely, more exotically. Remember Stanley Plog's research on personality and destinations. He was concerned with travelers, but the findings apply to travel journalists as well. Let your personality be your guide.

The outlet decision is the starting point in your approach to publishing. Think about where and how you want your work to be published. Then think about whom you will approach. More about this in Chapter 5, but you need a starting point, an outlet for your first work. Control is a key issue. If you want to exercise the greatest amount of control, start a travel-related blog and publish there. If you are willing to cede control, submit the work to an editor or publisher.

Cost is part of the decision-making matrix in two ways. First, the audience for travel journalism is price-conscious. Cost is a key factor in its decisions about where and how to travel.[109] Travel journalism reflects this: After "secret" and "hidden," the most common words on travel magazine covers are "you can afford." Some of my travel journalism, for example, is aimed at the frugal travel audience.[110] These aren't dilettantes: They know the price and the value of things. I was introduced to the concept of frugal travel by Arthur Frommer's *England on 60 Dollars a Day*. For more than two decades, I tromped around London following Frommer's advice. Later, Sandra Gustafson's *Cheap Eats in London* and *Cheap Sleeps in London* proved that fine yet inexpensive food and lodging were possible in this famously expensive city. The *Frugal Traveler* column in the *New York Times* often has excellent advice. Most recently, James Sherman's "Smart Luxury Values" suggested ways to find luxury for less.

Cost is a part of the decision-making matrix in a second way: What are your costs to travel and who will pay them? Funding travel journalism is covered in Chapter 7. For now, all that's important is to recognize that travel is relatively expensive and that it is unlikely that beginning travel journalists will be reimbursed. Another reason, perhaps, to start close to home.

## Test Your Understanding

1. Travel journalists depend on a matrix-like decision-making process that considers five choices. What are they?
2. Are travel journalists well aligned about the relative importance to readers and the frequency of coverage of the different types of travel stories? What does the survey research show?
3. Twelve topic niches vary in importance to readers, according to survey data. Select one of these topic niches and develop story ideas relevant to your market, or to where you want to travel.

## Practice Your Skills

Assignment #3—Develop a story idea consistent with what you've learned about your market in Assignments #1 and #2. Select from among the story types and, if appropriate, the topic niches. Draft a brief story memo—five to seven sentences—that answers these questions:

- What is the story type, the topic niche?
- Who is the audience?
- What is your story hypothesis, that is, what do you think the story will say?
- How will you test the hypothesis, that is, what is your reporting road map?
- What are the opportunities to tell this story in print and online?

# · 4 ·

# WHAT ARE THE ESSENTIAL SKILLS AND KNOWLEDGE FOR THE TRAVEL JOURNALIST?

The best travel book written by an American is Henry David Thoreau's *Walden* (1854), writes William Zinsser, "though the hermit of Concord hardly got beyond the town limits."[1] What Thoreau understood intuitively is, today, a studied practice of the travel journalist: doing important travel stories without the travel.

Indeed, Zinsser writes, this approach requires courage because the travel journalist working locally "relinquish[es] all the benefits a writer gets just by stepping off a plane in a foreign country: instant otherness."[2] Instead, the travel journalist working locally must find the important and interesting angles nearby.

Lots of travel journalists applaud this approach.

The freelance travel journalists I surveyed offered four reasons for working locally. First, it's a good place to start, with few costs other than your time. Second, where you live is a destination for someone else. "Almost everyone's 'backyard' is a destination to someone else," one freelancer said. Third, working local isn't just a starting point, but also a longer-term opportunity to develop expertise, a reputation, a niche. And fourth, so-called "regional travel stories" are quite saleable. "Write about where you live," another freelancer said. "People travel there."

Certainly, there are important and interesting travel stories to be done in most markets. They touch all four story types and many of the topic niches. Here are five approaches, drawn from work observed across the U.S.

## Cover the "Misery" of Air Travel

Michelle Higgins, the *Practical Traveler* columnist for the *New York Times*, was writing about the VIP treatment American Airlines offers ordinary, coach travelers—for a fee. "Airlines say they are offering such services," Higgins wrote, "as a way to address the aggravations of flying—particularly at the airport itself."[3]

Her story likely resonated with lots of air travel readers.

Fifty percent of U.S. adults travel by air each year for business or leisure[4]—this represents a large audience eager for news, enterprise, and investigative reporting. It's also a largely unhappy audience. Crowded airplanes, weather disruptions, connecting flights, misplaced luggage, intrusive security checks, departure delays, stingy frequent flier programs, and runway strandings add up to "The Misery of Flying," as *The Economist* called it.[5] However unhappy, these 206 million U.S. adults continue to depart from and arrive at the nation's 376 airports with scheduled airline service.[6] Busy, densely populated states such as California have as many as 21 airports. Even sparsely populated states such as Wyoming have two. No other country has as many.[7] State and local governments spend $1 billion a month for terminals, runways, towers, and other facilities.[8] Atlanta's Hartsfield-Jackson Airport, the nation's busiest by passenger volume, spent $1.28 billion to add a single runway.[9] The benefits to air travelers from airline deregulation have amounted to billions of dollars,[10] but also resulted in ever-changing, hard-to-understand fares, fees, and taxes. Airlines, since deregulation, go in and out of business, and in and out of markets.

Cultivating sources is one key to covering the "misery" of air travel. Get to know the people in your market who are the first to know when problems occur. Travel planners in large companies, government offices, and universities. Commercial vendors in airports. Law enforcement and other first responders. Airport managers.

Another key is government data, largely supplied by the Federal Aviation Administration (FAA). As an example, the FAA's website displays up-to-the-moment flight delays, by region or airport. As another example, the site displays preliminary and final reports on accidents and incidents, large and small, including links to related reports of the National Transportation Safety Board. The two types of information are always on the FAA's "top requests" list.[11] A third key is anticipating events that strain air travel, especially weather and holidays.

And, it also means looking for stories that are antidotes to the "misery" of air travel—as Higgins did for the *New York Times*.

## Mine Opportunities for Local "Watchdog" Coverage

Travel and tourism present an excellent opportunity for "watchdog" coverage, according to travel editors I surveyed. Among those who responded, 59 percent somewhat or strongly agreed. Same for the business writers and editors I surveyed: Among those who responded, 70 percent somewhat or strongly agreed.[12] Two opportunities are common.

First, most markets are served by a destination marketing organization, usually called a convention and visitor bureau (CVB). Most CVBs are independent nonprofits. Some are government agencies. A few are divisions of the local chamber of commerce.[13] Their task is to promote travel and tourism as a function of economic development.

The opportunity for "watchdog" coverage has to do with the way CVBs are funded. Most are funded by an array of "visitor taxes," including room taxes, rental car taxes, and restaurant taxes. The average annual take of a CVB is $4.8 million.[14] CVBs spend 48 percent of their budgets on sales and marketing, 39 percent on personnel, and the rest on administration. The first question is: How effectively are these taxpayer funds spent? Most CVBs commission third parties to calculate the "economic impact" of travel and tourism in their market. Journalists who cover CVBs say it's difficult to substantiate whether these calculations are accurate,[15] but they are helped by the fact that CVB records— including the studies of "economic impact"—are accessible as public records.

The second opportunity for "watchdog" coverage has to do with the way seemingly private tourist attractions can become public albatrosses. All manner of attractions—most of them tied to arts, culture, or history—come to rely on taxpayer subsidies for their survival. Consider recent examples in the state of Georgia, a $21 billion tourism market.[16] In one Georgia market a Civil War naval museum opened in 2001 was expected to attract 160,000 tourists annually and thus would need an $80,000 taxpayer subsidy only for its first year of operation. Ten years later, it attracts just 20,000 tourists annually and its taxpayer subsidy has grown to $300,000.[17] In another Georgia market, a 15-year-old, $6.6 million music hall of fame closed in 2011 because local taxpayers could no longer subsidize million-dollar annual losses.[18] In a third Georgia market, a $14 million fishing museum became what the *New York Times* called "a symbol of waste" in a state struggling to pay its bills.[19] Can you identify one or more attractions in your market that will fail without an unexpected taxpayer subsidy?

## Find the "Hidden," "Secret" Attractions That "Nobody Knows"

This approach is, well, fun. Everything interesting in a market is known to someone. But many interesting things are known to just a few. Guides almost everywhere make a business of showing travelers the "hidden attractions," the "secret places," the venues that "nobody knows." Do the same for your market: Find the "Plymouth Rock" for modern-day immigrants. Introduce a strange grave in an obscure cemetery. Locate the earliest genealogical record in the public library. Find the tree from which someone was lynched. And so forth.

This approach is as instructive as it is fun. You'll practice all of the story sourcing skills essential to the travel journalist: interviewing, secondary sources, documents and data, observation. You'll gain knowledge of the market that will contribute to expertise, reputation, and niche.

Publish in print and online. Try it as a Service and Advice story. Then a Destination story. And then a Journey.

Even offer to conduct tours.

## Put Your Market in Context

One approach is to show how a market is unique. Another is to show how it's not. An interesting venue in your market may be one of a kind, but probably not. The museum-like doll shop in your historic district? There are 51 others across the U.S.[20] The family-owned, award-winning barbeque joint that features a pork chop sandwich with a mustard-based sauce? Nope: more than 10,000 hits on a Google search to sort through. How about a condom manufacturing plant? Ninety to 100 others.[21]

One opportunity, as discussed in the approach above, is to report about this venue in the context of your market as a destination: If you come here, visit this. But the additional opportunity is to place this venue in a broader context. Any others like it in the state, region, nation, world? Any others related to it in some way? The further opportunity is to report about this in such a way as to build a network, a community of interested travelers.

Try this experiment: Think of an aspect of your market that seems to be one of a kind. In the area where I live for part of the year, for example, is the house where Ma Rainey, the "Mother of the Blues," was born. Today, it's a slow-developing museum. But it is a venue associated with an important American musician. Might there be other Ma Rainey–associated venues anywhere else in the U.S.? There are no other museums, per se. But a Google search on keyword "Ma Rainey" pulls up 431,000 sites. Scroll through and a long list of venues turns up: the studio in Chicago where she first recorded, the string of clubs where she famously toured the "Black Bottom," a shop where her sheet music is still sold, her display at the Rock and Roll Hall of Fame, the theater where August Wilson's play about her life debuted, the postal museum that houses the original artwork of the 1994 Ma Rainey commemorative stamp, her grave in Rome, Georgia, and so on. The opportunity in this example is to build a travel-related site that links these venues. I'd start, of course, with the venue in my market. It's what I know. Add from there, reporting from a distance. Use social media tools to invite other travel journalists to add to the site. Build a community of interest around Ma Rainey and the locations where Ma Rainey–associated venues can be seen. What's a comparable venue in your market? Pick one and follow the process. Is there a travel journalism opportunity?

The Ma Rainey example links a venue in one market with related venues elsewhere. Another approach is to link markets demographically. Consider, for example, the work of Patchwork Nation,[22] a foundation-funded journalism project. Patchwork has grouped the nation's 1,361 counties by "community type," based on demographic factors such as income, race, employment, and religion. Journalists use these groupings to shape their reporting about these counties. How might the Patchwork Nation approach be useful to a travel journalist? At the simplest level, it's one data point about a market: Visit Cleveland, Tennessee—one of the nation's "evangelical epicenters" (one of the 12 "community types"). The larger opportunity is to build a network, to form a community of travelers and travel journalists interested in, in this case, "evangelical epicenters."

A third approach, perhaps the most common, is to link markets based on similar public events such as festivals. A fourth approach is to link markets according to issues, with attention to, for example, protest sites.

## Debrief Travelers to and from Newsworthy Places

Travelers from your market might find themselves in interesting places when news happens. It's common journalistic practice to interview such eyewitnesses to history. But there is a larger, deeper opportunity. Residents in your market travel outside the U.S. for many reasons other than tourism. They are members of the military on deployment, in war zones as well as other hot spots. They are medical providers on missions, often to places that have experienced disasters. They are college professors conducting research, sometimes with public policy implications, on issues such as climate change or oil spills. They are business executives employed by global firms, often in the ferment of global financial activity. They are students studying abroad.

The opportunity is to tell their story, to see the world through the lens of their experiences. Unlike the hapless tourist caught up in the news, they are away intentionally, for longer periods, bringing some level of expertise to bear. Often they live on the local economy. Some have language skills. But most importantly, they are local residents with whom your readers and viewers can connect. Their foreign story becomes a local story.

There are many ways to approach this story:

- An academician in Cedar Falls, Iowa, produced a television program called *Here and There* on the public access channel, with interviews with local residents returning from newsworthy areas.[23]
- Local residents traveling abroad were encouraged to write a "Letter Home" for the *Sentinel* in Keene, New Hampshire. Often, they were interviewed upon their return.[24]
- Q&As are a popular debriefing tool. A good example is *Conde Nast Traveller's* column *The Conversation/The Global Citizen*. A brief bio and set of travel tips accompany the questions and answers.[25]

Similarly, interview the same sorts of people who come *to* your market from foreign countries.

## Other Approaches

- Adopt the role of ombudsman for travelers in your market. There are lots of models to emulate: Christopher Elliott, the reader advocate for *National Geographic Traveler*, also syndicates a weekly column to newspapers;[26] the *Miami Herald* answers travelers' questions as a part of its consumer *Action Line* column;[27] *USA Today* publishes a *Traveler's Aide* column written by Linda Burbank.[28]
- Be aware of opportunities to cover public-policy disputes involving travel and tourism. As an example, tourism marketers have generally opposed state legislation tightening rules on immigration.[29] Such rules often result in travel boycotts that damage the travel and tourism industry.[30] Two opportunities: Cover the debate in terms of the travel and tourism angle, or determine, a year after the new rules, whether travel and tourism were, in fact, impacted.

## Travel Journalism on the Road

Two imperatives influence travel journalism on the road. One is journalistic excellence. The other is journalistic efficiency. In other words, get as much good stuff as possible for the least investment of time and money.

## The "Three-by-Three" Model

The "three-by-three" model is an effective way to accomplish both. The model requires that the travel journalist report, write, and publish (or prepare to publish) three stories over a three-week cycle.

The three stories are:

- Story #1: The "sure thing" story. You know it's there to be gotten. And that this story, by itself, is worth the time and money. You either have an assignment or you know, from experience, that it is feasible to produce "on spec." Nothing foreseeable will keep you from getting this story. As American Express used to say, in a related context, "Don't leave home without it."
- Story #2: The story that results from serendipity. Serendipity, which means making fortunate discoveries accidentally, is the travel journalist's "unheralded

goddess."[31] On the track of one story, you stumble upon another. The key, of course, is a journalistic frame of mind that is aware and open to this opportunity. Charleston *Post and Courier* reporter Tony Bartelme went to Shenzhen, China in pursuit of one local angle—and returned with four. "This turned out to be a surprisingly easy task," said Bartelme.[32]

- Story #3: The story that results from "gathering string" around a travel-related topic. To journalists, gathering string means collecting bits and pieces that, once put together, "reveal something new or capture an interesting slice of life."[33] Keep a notebook labeled "Things I've Done Wrong," for example, or "Things I've Done Right."[34]

The three-week cycle operates as follows:

- Week #1: At-home pre-work. "Success . . . isn't about what transpires in a foreign land," advises Michael J. Jordan, "it lies in the thorough preparation done back home, well before your departure."[35] With a saleable idea and an outlet in hand, Jordan and others recommend these pre-work steps: Mine secondary sources for background and context, consult trade associations for the names of experts, set an itinerary for the first few days, and contract with guides and translators.
- Week #2: The reporting trip. The idea is to show, not tell, the story. To do this, you have to see the story in play. That's the essence of the reporting trip. Find the story's places, find the story's people. Observe them, record them. Show the story through them.
- Week #3, Back-at-home writing and publishing (or preparing to publish). You should have the "tell" parts of the story from the pre-work, the "show" parts of the story from the reporting trip. You should be structuring the story in your head (and your notebook) as you go. Now, write what Anne Lamott calls "the shitty rough draft"[36] quickly, then revise.

Here's an example, from my experience, of the "three-by-three" model at work:

The week of pre-work convinced me that I could produce a strong package of ecotourism-oriented action narratives in the Monteverde area of Costa Rica. Costa Rica is the most visited ecotourism destination in the Americas and one of the three most visited in the world.[37] Monteverde's cloud forest reserve is the most important wildlife refuge in the Americas.[38] Its experimental farming cooperative, Finca la Bella, is a compelling example of reforestation, sustainable agriculture, and agrarian land reform.[39] This area, in north central Costa Rica, is accessible enough for a quick, one-week trip, but remote enough that it isn't as known to travelers to Costa Rica as the Arenal Volcano, Drake Bat, or Manuel Antonia National Park.

Again, based on pre-work, I knew the Monteverde area was developed by Quakers from my region of the U.S. I also knew that my university operates a research station and an eco-lodge in Monteverde, so I had a pair of local-global angles that made the story sellable, experts on sites, and good access to guides and translators.

At the rate of one action narrative per day, a package of five for the week was gettable, my "sure thing." I'd stay alert for serendipity and gather string around local food.

The week, as planned, produced five action narratives:

- A trek to the San Luis Waterfall
- A cheese factory with an unexpected history
- Small-scale coffee with a high-tech twist
- In search of the elusive resplendent quetzal
- A last glimpse of wildlife

Serendipity bit me, so to speak, the last few days in Monteverde. Here's what I wrote:

Room and board for four nights at the University of Georgia's Ecolodge San Luis: $250
Treatment for the black fly bites: $500
 Seems about right, given my own stupidity.
 The "what to bring" list was explicit: long-sleeve shirts as a barrier to biting insects.
 I brought short-sleeve shirts, exposing my arms just above the elbow. But, no problem, I calculated. I've brought a 100 percent DEET-based insect repellant, sure to ward off all biting insects.
 Mosquitos, si; black flies, no.
 In fact, according to a University of Florida researcher, DEET-based insect repellant actually attracts black flies.
 And attract it did.
 By visit end, I had nearly 20 black fly bites. They surrounded the elbow joint on both arms, the welts rising, like tribal markings, or crude prison tattoos. They itched beyond distraction, sometimes disrupting sleep.
 Upon my return, I consulted my dermatologist, who prescribed two antibiotics—one costing $248—a topical cream, and a soaking bath.
 The welts and the itching went away within a couple of days.
 Lesson learned: long sleeves in the tropics.

As to gathering string, I decided to begin collecting the experiences of cooking lessons in foreign cultures. Marina Zamora Elizondo, one of the cooks at Ecolodge San Luis, agreed to teach me.

I drafted, revised and published the stories a week later at a simple-to-construct, free website on the Google Sites platform.[40]

# Reporting and Writing Skills in Five Areas: Sourcing, Elements, Structure, Drafting, and Revision

The dominant how-to books on travel writing are filled with advice about crafting travel stories. I won't repeat that advice here. I offer, instead, advice about sourcing, elements, structure, drafting, and revision that contributes to this book's purpose: a more journalistic approach to travel writing.

# Story Sourcing

It's often observed that journalists depend too little on observation—their senses of sight, smell, touch, taste, and hearing. The opposite could be said of travel writers: They depend too much on observation.

The travel journalist ought to use all five of the sourcing methods common to journalism. They are:

Interviews—Interviewing is "at the heart of most nonfiction reporting," James B. Stewart has observed.[41] Yet, all too often it plays no part in travel writing. Interviews lead everywhere: to secondary sources, data, documents, and opportunities to observe. Prepare well so as not to appear foolish and to increase the chances of getting what you want. Asking "why" requires an explanation. Asking "how" requires a description. Answers contribute to voice. Language barrier stand in the way? It need not. Translators are available almost everywhere. More on this in Tips, Tools, Cautions below.

Secondary sources—Getting smart about the travel topic, the topic niche, is part of at-home pre-work. Find secondary sources that provide the "essential one inch of reading," as an editor whose name I've long forgotten once advised me. Topic-specific periodicals and trade association newsletters often are the best sources. Search online in the *Standard Periodical Directory*, the *Oxbridge Directory of Newsletters*, or the *Encyclopedia of Associations*. Use what you learn in your story as background and context, signaling readers about how things work, how parts connect, how things matter.

Data—The journalist as "math-phobic" is a foolish—indeed dangerous—deprecation. All serious people—including travel journalists—embrace math as an invaluable tool for understanding everything. Here's a way forward: Think of data as information. Think of numbers as language. Looking for a role model? Read Bill Bryson's *In a Sunburned Country* (2001). Bryson uses data 45 times to rank, rate, record, or index Australia—in the first six pages. No one would accuse humorist Bryson of dense, turgid prose.

Documents—Journalists regard documents as the "best evidence" because they are contemporary, part of a cultural trait to "record" human activity, and accountable. Documents are most relevant to the travel journalist when covering travel and tourism industry news. Consider, for example, tax records when the subject is convention and visitor bureaus and how they spend hotel/motel tax receipts. Safety records when the traveling amusement park arrives in town. Court records when the topic is sex tourism.

Observation—Observation takes advantage of the five senses: sight, smell, touch, taste, and hearing. As noted earlier, travel writers depend too much on observation. But there are three areas where observation is an essential sourcing method: in describing settings and characters; in gathering anecdotes; and in recording dialogue.

Travel journalists also should take advantage of the newest thinking about audience participation and the technology that facilitates it. Simply put: Make the audience for travel journalism an effective participant in framing stories and gathering information.

"Citizens everywhere are getting together via the Internet in unprecedented ways to set the agenda for news, to inform each other about hyperlocal and global issues, and to create new services in a connected, always-on society," write Shayne Bowman and Chris Willis. "The audience is now an active, important participant in the creation and dissemination of news and information, with or without the help of mainstream news media."[42]

How might audiences participate?

"The readers can give us facts we did not know," writes Dan Gillmor. "They can add nuance. They can ask follow-up questions. And, of course, they can tell us when we are wrong."[43]

At the simplest level, the travel journalist needs to maintain a systematic network of sources she regularly pulses. These are the local market military and civilians on deployment, health care providers on missions, faculty conducting research, executives on overseas assignments, expatriate retirees, students studying abroad. Reach them through a listserv or a social media site. Or adopt the more sophisticated approach of Minnesota Public Radio's *Public Insight Journalism*, which harnesses a network of thousands of sources who share expertise and experiences.[44]

Beyond U.S. borders is help from Global Voices, an online aggregator of what cofounder Ethan Zuckerman labeled "bridgeblogs." These are blogs, Zuckerman writes, "that reach across gaps of language, culture and nationality to enable communication between individuals in different parts of the world."[45]

They are useful to journalists, Zuckerman says, as another tool to report what people are saying about global events. Managing editor Solana Larsen says journalists increasingly turn to Global Voices to contact bloggers who can provide insight from a country that's in the news.[46]

Technology facilitates audience participation in new and fascinating ways.

Search-engine data can help travel journalists identify and document trends. How does this work? Researchers have discovered that they can tap into the public's mood, interests, and behavior by tracking what the public searches for online.[47] Similarly, travel journalists can use social media to find sources, raise questions, get answers.

The key is doing so in a cautious, systematic—that is to say, journalistic—way.

Search-engine data, for example, helped journalists and others track the outbreak of swine flu in 2009 and the emergence of the Arab unrest in 2011.[48] Data from millions of searches amounted to "a very broad poll," one researcher concluded.[49] Reporter Paul Lewis used the microblogging tool Twitter in 2009 to identify 20 witnesses to the beating death of a bystander at the G20 meetings in London.[50] Poynter's Roy Peter Clark used open-ended questions about women and high heels on his Facebook page and got back "possible sources, story angles, themes, tensions, links and leads."[51]

Tools abound. Dozens of software developers are producing tools to track use and glean insights from Google, Facebook, and Twitter. Just three years ago, for example, five research tools for Twitter dominated: TweetStats, Trendrr, Tweetmeme, Twellow and Xefer Twitter Charts.[52] Today there are hundreds, including Twitter Search and Twitter Scan, probably the most useful pair for travel journalists. The same kinds of tools are available for Google and Facebook.

Caution is urged by journalists experienced with this approach, which is called crowd sourcing. "The pitfalls of crowd-sourcing keep me awake at night sometimes," Lewis says. Among his concerns: misinformation, propaganda, and difficulty with verification.[53]

Travel journalists show some evidence of adopting these approaches:

- Rebecca Sebek recommends using the Google Keyword Tool and Google Trends to identify untapped travel writing niches. Three niches her research identified: weight loss travel, meditation travel, and divorce travel. "With a little research," Sebek writes, "you could stumble upon an untapped market that will catapult you to the top."[54]
- TheCityTraveler.com used Twitter to link readers with a mobile-location-loyalty application called "foodspotting"[55] in the hunt for outlets in 29 states selling Victory Beer's Summer Love Ale.
- TheVacationGals.com invites "guest blog posts from moms, dads, frequent travelers, book authors and others who have vacation-related insight that would be of interest to our readers." The site uses Facebook to solicit "thumbs up" for a list of top family travel blogs. It poses questions such as "Who's traveling this weekend?"
- Ian Sluder, an authority on Belize, invites reader questions and offers free answers on his website at http://www.belizefirst.com.

But the potential is far greater: Fewer than 7 percent of the blogs and websites run by the freelance travel writers who responded to my survey showed any use of crowd-sourcing tools.[56] That said, far more use social media tools—Twitter, Facebook, Google+—and various comments applications[57] to engage and interact with readers, which perhaps is a starting point.

## Story Elements

Topic and structure influence—even dictate—story elements. But it is safe to say that regardless of topic or structure, these elements will appear in some form and place in all travel stories.

Lead—Some writers, according to Roy Peter Clark and Don Fry, spend as much as 50 percent of writing time on leads.[58] Others dash them off at the end. Whenever written, all leads hope to accomplish the same thing: attract a reader's attention. Lots of advice exists about writing leads. Here's advice especially applicable to travel journalism:

- James B. Stewart says the lead must attract and hold readers by re-creating in their minds the same curiosity that drove you to undertake the story.[59]
- David Mehegan says he tries to do in the lead what he tries to do throughout the story: always speak directly to the reader's desire for interest and enjoyment.[60]
- Kevin Cullen says the lead should set the tone for the rest of the story. If it's a bizarre story, use a bizarre lead. If a sad story, a sad lead. If funny, funny.[61]

Nut graf—If the lead's goal is to attract, the nut graf's goal is to explain and promise. As William Blundell notes, in the lead it was "tease me, you devil"; in the nut graf, it's "tell me what you're up to."[62] Clark and Fry describe the nut graf's explanatory function with this three-way test:[63]

- What the story is about
- A key fact the reader needs to understand the piece
- Why the reader should read the story

Background and context—Background reminds the reader of what has gone before, says Graham Watts.[64] Context places the story within its surroundings. Both are essential roles of journalism. Here are three seed-phrases that help frame background and context:

- XXX is not alone . . .
- Across the state, cities like XXX . . .
- Since 19XX . . .

Body—However structured, the body of the story contains the proof. The lead teases, the nut graf explains and promises, the body delivers. Many journalists "audit" their leads and nut grafs after writing the body. Did I deliver what I promised? If not, add proof to the body or recast the top.

Ending—Endings, according to James B. Stewart, tend to accomplish one or more of the following:

- Resolve unanswered questions
- Sum up the significance of the story and suggest its ramifications
- Inspire further reflection and contemplation[65]

# Story Structure

"All good stories have a structure," says James B. Stewart, "which unifies even seemingly disparate elements."[66] The story structure answers the writer's question, Where do I go next? And the story structure reassures readers that there is a logic to the story.

For the travel journalist, story type influences story structure.

News-type travel stories follow the same structure as news, enterprise, and investigative stories on any beat. The simplest news story might well follow the inverted pyramid structure. But as the story becomes more complex—if it contains strong findings or assertions of wrongdoing, for example—you need a different structure, one associated with "watchdog" journalism.

**Watchdog**

| | |
|---|---|
| | Anecdotal lead: Typically, the victim and the bad guy |
| | Nut graf: Generalizes the scope and scale (i.e., "Victim is not alone. Each year, 10,000 people just like victim…" |
| Bulleted findings or assertions (i.e. "A two-month inquiry indicates that ..." ☐ ☐ ☐ | |
| | Background and context |
| | Proof of findings or assertions |

Service and Advice-type travel stories follow the same structure as Service and Advice stories on any "consumer journalism" beat. Often they are Q&As, lists, and "if you go" boxes. The key is simple, clear, direct prose that answers the reader's questions just as they occur to the reader. Examples:

- We found the three cleanest, safest, cheapest backpacker hotels in Rome
- Here's the way around your airline's baggage-handling charges
- London on $100 a day—including hotel!

More complex Service and Advice-type stories amount to trend stories and should be structured accordingly.

Destination and Journey stories are structured similarly. They contain both narrative and analysis or, in other terms, anecdote and meaning. In the narrative elements, the writer takes the reader into the action, showing. In the analysis elements, the writer pulls back, telling the reader what the action means.

Roy Peter Clark, citing the work of Nicholas Lemann, proposes a structure for this type of story: broken-line narrative.

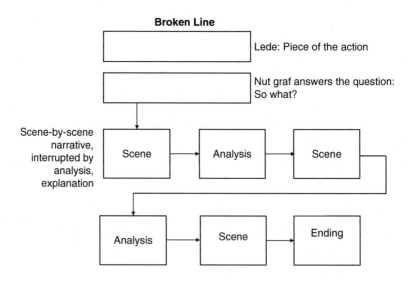

# Drafting and Revision

I urge writers to adopt a two-step process that emphasizes fast drafting and methodical revision. One influence is Anne Lamott, whose *Bird by Bird* is embraced by writers for its "instructions on writing and life."[67]

The chapter "Shitty First Drafts" makes the argument: "For me, and most of the other writers I know, writing is not rapturous. In fact, the only way I can get anything written at all is to write really, really shitty first drafts."

Get a draft. Get past that barrier. Then revise.

"All writers write badly—at first," says Donald Murray. "Then they rewrite."[68]

A fine practitioner of the art of revision is Poynter's Roy Peter Clark. His *Writing Tools: 50 Essential Strategies for Every Writer* contains invaluable advice. Unlike Strunk and White's *Elements of Style*, nothing is vague or philosophical in Clark's work.

The Clark revising tools I commend to all writers, in particular travel journalists, are Tools #1–10 under "Nuts and Bolts" in the first part of the book. Start with a draft of your story and a single, indispensable tool: a highlighter.

Let's just take one of the tools as an example—the first one, "Begin sentences with subjects and verbs." Clark's advice is threefold:

- Compose a sentence with subject and verb at the beginning . . .
- . . . followed by other subordinate elements . . .
- . . . creating what scholars call a right-branching sentence

So, how to apply the tool? It's a three-step process: highlight, ask, revise:

- Take a draft of your story
- Highlight the subjects and verbs
- Ask: Does the sentence follow the rule?
- Revise accordingly

# Tips, Tools, and Cautions

Here's a set of tips, tools, and cautions, drawn from many sources, which should help you cover travel and tourism in your market or on the road. These should be of especial help to travel journalists seeking more substantive, independent foreign correspondent-type stories.[69]

## Bridging Cultural Divides

Marda Dunskey, a journalism instructor at DePaul University, suggests five criteria to guide reporting about another culture.[70] They are:

- Balance: The range and mix of sources
- Point of view: From whose perspective is the story told?
- Voice: Who is quoted? Who gets to speak?
- Context: What are the relevant historical, political, cultural factors?
- Framing: Which issues are included, omitted?

A highly regarded source of primer information on all of the world's religions is at http://www.beliefnet.com. Follow the navigation under "Faiths & Practices." Recently, the site was honored with an award for general excellence by the Online Journalism Association (its top award).

# Consult Tip Sheets, Story Archives, Databases, and Think-tank Backgrounders for Ideas and How-to Advice

Tip sheets—Start with the Travel Writers Exchange,[71] a gathering of travel writers, bloggers, and journalists. "We exchange tools, tips, and resources," the site's Trisha Miller writes, "and provide a forum to help our community expand their online reach and find new opportunities."[72] A robust search tool helps you find what you need. One of the best sites for tips about publishing travel journalism online is the Writers' Website Planner created by veteran guidebook author Tom Brosnahan[73] Brosnahan's advice about blogs versus websites is particularly helpful. On the "watchdog" side is Investigative Reporters and Editors (IRE) and its tip-sheet database. The database includes 3,000 tip sheets gathered from presenters at IRE conferences. A search for tip sheets on "travel," for example, produced 15 tip sheets dating from 1990 to 2010.[74]

Story archives—No single archive exists for best-practice examples of travel journalism. Award-winning stories often trigger ideas and suggest how-to advice. The first place to look is the list of winners of the Lowell Thomas Awards.[75] The winners dating back to 2008 are listed online—hundreds of stories, photographs, guidebooks, websites, and videos in 25 categories. A list of winners from earlier years is available by mail from the Society of American Travel Writers Foundation.[76] Most, but not all, of the winning stories are findable online.

Databases—Over time, you'll work with databases from many sources. Get started with U.S. Census databases by learning how to use its American FactFinder tool. There are excellent tutorials online.[77] Use SocialExplorer.com to map U.S. Census data and Tableau Public to visualize data in many forms. They are free.[78] Virtually every other federal database open to the public is at FedStats.[79] Audience data, as discussed in Chapter 2, is from SRDS's *Lifestyle Market Analyst*.[80]

Think-tank backgrounders—One of the best ways to stay current about global issues is to consult the Council on Foreign Relations. Its "crisis guides"[81] won the 2007 Knight-Batten Award for innovation in journalism. Also, http://www.alertnet.org and http://www.reliefweb.int offer backgrounders on humanitarian crises.

# Finding "Fixers" for Translation, Logistics

Travel journalists often depend on "fixers" for translation and logistics. This is especially true for "parachutists," the travel journalists on temporary assignments outside the U.S.

Finding a fixer before leaving the U.S. is relatively easy—and relatively cheap. Local freelance writers, college English teachers, and independent tour agency guides are good sources.

Researchers Jerry Palmer and Victoria Fontan, writing in Sage's *Journalism*, cite three "risks that arise from lack of language competence and dependence upon . . . fixers." They are:

- Mistranslation and/or omission of significant material
- Inability to "blend in" with the local population and to understand the local culture
- The fixer "forming" the journalist's view of the situation[82]

The travel specialists Wendy Perrin touts as "Perrin's People"[83] are a good starting point. Her first choice of expert for Cambodia, Laos, and Thailand, Andrea Ross, arranged interpreters for me on travel journalism assignments in Bangkok, Phnom Penh, and Hanoi—all first-rate and none more than $50 a day.

But do choose carefully: Ulf Hannerz tells the story of journalist Vincent Dahlback, reporting in Rwanda, who hired a Hutu fixer and a Tutsi driver. Bad mix, he quickly discovered.[84]

## Travel for One Story; Come Back with Four

Tony Bartelme, a reporter for the *Post and Courier* in Charleston, South Carolina, followed a local man to China where he was to have stem cell treatment.

While there, Bartelme and photographer Alan Hawes tried to find other "local" stories.

"This turned out to be a surprisingly easy task," Bartelme wrote in *Nieman Reports*.[85]

Some examples:

- They visited a neighborhood in Shenzhen where 3,000 artists churn out fake van Goghs and Rembrandts in an assembly-line manner. "This framed a story about how cheap knockoffs affect Charleston's art scene," Bartelme wrote.
- They toured a factory owned by a Charleston-based lighting manufacturer. "Our tour of the factory helped to shape an in-depth story about how China's boom directly affects businesses, employment and consumers in South Carolina," Bartelme wrote.
- They visited a port that's learned how to squeeze containerized cargo into a place where land is scarce. "This issue is of great importance to Charleston," Bartelme wrote, "which has a large container port and is struggling to find new space to expand."

## Learn Cheap, Simple, Multimedia Tools

Cost and complexity no longer stand in the way of producing high-quality multimedia travel journalism.

Mark E. Johnson, a lecturer at the University of Georgia's Grady College of Journalism and Mass Communication, tells photojournalists that they can begin producing audio slideshows *and* video for as little as $295.

Given the digital camera and Mac or PC already in their tool kit, here's what Johnson tells them to buy:

Hardware:

- Olympus WS-310M digital recorder ($100)
- Nady SP-4C mic ($10)
- Radio-Shack adapter ($5)
- Flip Video camera ($130 for 30 minutes, $150 for 60 minutes)

Software:

- EasyWMA (Mac only, look for Switch for PCs) ($10)
- Audacity (don't forget the required LAME encoder, under Optional Downloads) (Free)
- SoundSlides ($40 or $70, depending on the version you prefer)

# Translate from English to Another Language to Expand Reach

Have a story you've written in English that might appeal to readers of other languages? Consider expanding your story's reach by collaborating with one or more of the fastest growing news sources in the U.S.—ethnic media.

New American Media, a collaboration of ethnic news organizations, produces the *National Directory of Ethnic Media* which provides information on more than 2,000 ethnic media organizations in the U.S., including print, online, radio, and television. The print edition is organized by ethnicity and media type, with alphabetical and geographical indexes. The online[86] edition also includes a custom search feature that allows you to create your own search for media by city, state, language, ethnicity, and media type.

# When Traveling, Consult Local English-language Press

The English-language press is widespread throughout the world. Any country colonized by the British Empire is likely to have an English-language daily newspaper. They are excellent sources to consult here, or when traveling.

Find a list at MediaChannel's Global News Index, at http://www.mediachannel.org.

You'll find these newsrooms open and inviting. Indeed, some beginning journalists seek work in English-language dailies. "You get to know the country and the players," says Caitlin Randall. "You make your contacts. You also get clips, which show that you have written in that country."[87]

One caution: "Lifting" from the English-language press in foreign countries—that is, using their reporting without attribution—is both common and wrong. Better to attribute, showing that you understand the value of their reporting.

## Aid Agencies as "Sources, Gatekeepers, Eye Openers"

The UN's Hilaire Avril summarizes the symbiotic relationship between aid agencies and journalists this way: Aid agencies "rely on journalists to raise awareness" of their causes and "to attract donor funding," he writes in *Nieman Reports*, and journalists "rely on aid workers to gain access" to their stories. Yet, both sides are wary of each other's motives and behavior, Avril says, and should be.

A study by Columbia University's Steven Ross contains advice for journalists covering humanitarian crises.[88] Among the simplest and best: Make better use of the two best sources of information on humanitarian crises. They are http://www.alertnet.org, which is a product of the Reuters Foundation, and http://www.reliefweb.int, sponsored by the UN's Office for the Coordination of Humanitarian Affairs.

Jumo.com, created by one of the founders of Facebook, is a good tool for locating NGOs associated with a place or issue.[89]

## Nine Elements Missing from the Conventional Travel Story

These nine elements are listed in a tip sheet Thomas Swick distributed at the 2009 Nieman Narrative Conference.[90] The list is best used, I think, as an auditing tool. How many of these elements are missing from your draft? Is there any way to incorporate them as part of the revision?

1.  A personal voice—So many travel stories sound the same. If you write honestly about your reactions to a place—what about it surprised you, amused you, depressed you, intrigued you, moved you?—you

will inject a personal voice into your story and in so doing, make it stand out.

2. A point of view—There is often nothing negative in the conventional travel story, and this bland rosiness, by ignoring the realities, not only makes for dull reading, but also does a disservice to both the subject and the readers.

3. A sense of the present—The traditional travel story focuses on the history of a place—the monuments, the museums. The challenge and thrill of travel writing is discovering what is happening now.

4. Imagination—Travel writers are sometimes accused of making things up; the best save their creativity for language that—through the use of similes, metaphors, allusions—brings a true experience to life.

5. Insight—The conventional travel story skims the surface. Now that almost every place has been photographed, the best travel writing goes beyond the descriptive into the analytical.

6. Continuity—A good travel story is more than just a collection of random impressions; it has a theme. Decide at the beginning what point you want to get across about the place, and then work your story around it.

7. Humor—Travel, the displacement from the familiar to the foreign, is inherently funny, but most travel stories are devoid of laughs.

8. Conversation—The inclusion of dialogue, or even long passages of recorded speech, gives a travel story some of the qualities of a short story while making it both personal and topical.

9. People—It is difficult to write knowledgeably, or movingly, about a place unless you have made contact with at least one local. It is not the sights but the people you meet that give your journey its uniqueness; they give you insight into the everyday life of the place; and—if you're lucky—they give you an emotional connection.

# Build Cooperative Relationships with Non-U.S. Journalists in Countries of Interest

Non-U.S. journalists are eager for cooperative relationships with U.S. journalists. I first learned this on a 1990 reporting trip to Zimbabwe. Local reporters, who felt intimidated by government-owned media, were eager to get their stories told, and they offered ideas and background, and introduced me to sources.[91] Later, in 2010, an editor in Torreon, Mexico, told me the same thing about his efforts to cover the violence associated with the drug war.[92] A good example of this idea at work is Haiti News Project's pen-pal service, which links U.S. and Haitian journalists who want to "share their experiences, practice their English [and French] and hone their reporting skills."[93] The service was launched in September 2010. A year later, 40 journalists had signed up and were corresponding.

# Be Safe as You Travel

There are two essential sources of pre-travel advice. One is the Traveler's Health channel at the Center for Disease Control and Prevention's website.[94] Click on "destination" and the countries you intend to visit for detailed information about health risks, vaccinations, and medications. The other is the U.S. State Department's travel-oriented website, where you'll find detailed country information and up-to-the-minute travel warnings (serious) and travel alerts (less serious).[95]

## Test Your Understanding

1.  Travel journalists favor the concept of "travel journalism without the travel." They typically cite four reasons for working locally. What are they?
2.  This chapter identifies five approaches to find important and interesting local travel stories. Which of the five approaches seems best for your market and interests?
3.  Imagine and describe a reporting trip you might take based on the "three-by-three" model.
4.  Describe three ways the audience might participate in the story you imagined and described in Question 3 above.
5.  Why is a nut graf essential in travel journalism stories? Provide three reasons.
6.  From the "tips, tools, and cautions," select three and describe why they would be especially useful to your work as a travel journalist.

## Practice Your Skills

Assignment #4: Outline a story based on the story idea you drafted for Assignment #3. Incorporate story sourcing, story elements, and story structure. How might you use interviews, secondary sources, data, documents, and observation? What is the story's nut graf? What story structure seems appropriate? Which elements are best for print, which for online?

# A Closer Look: On the Road with a Media Tour in Corinth, Mississippi

To understand how media tours work, meet Karon Warren.

Warren is a full-time freelance travel writer and photographer whose work appears primarily in newspapers. In 2010 she went on six expense-paid media tours. One was to a conference center in the north Georgia mountains. Another was a cruise on the Carnival line. A third was to a golf and tennis resort in Florida.

I followed up with Warren about a media tour to Corinth, Mississippi, sponsored by the local tourism bureau on the eve of the sesquicentennial of the U.S. Civil War. The tour organizer, a Florida travel PR firm, recruited eight writers including Warren to tour battlefields, visit museums, and dine on the local fare.

Warren sold two versions of the story, one to a newspaper in Pennsylvania,[96] another to an online destination guide.[97] The expense-paid tour cost the sponsor about $13,000—a little over $2,000 per travel writer.[98]

Only one of the six travel writers who responded to my survey disclosed to readers that their expenses were paid. Warren did not. Asked why, she said: "The media tour allowed me to see what was available for travelers to Corinth, but it did not affect what I wrote about my visit."

It sure seemed to. Warren wrote about the stop-by-stop venues she and the others visited over the three days. Same battlefields, same museums, same restaurants, same motels. The language she used to describe these venues— "another must see . . . no visit is complete without . . . a unique culinary treat . . . a vivid and notable history"—is characteristic of the media tour narrative.

Warren's Corinth trip is typical of how media tours operate. And the media tour model is typical of how a lot of travel writing gets produced and published. It's a model that is at odds with established journalism ethics. And more than anything else, it's what sets many travel writers apart from their journalism colleagues.

I surveyed freelance members of the Society of American Travel Writers. Some 65 percent who responded to my survey participated in one or more media tours in 2010.[99] For most, it was a trip or two, in the U.S. or abroad. One reported 10 domestic trips and three overseas trips. All but a handful accepted free travel, accommodations, and meals from their hosts.

Media tours come in all sizes and shapes. Most are for groups of travel writers. Some are arranged for individuals. Among the largest media tours are those organized in connection with regional, national, or international meetings of travel writers, including the Society of American Travel Writers. The International Food, Wine and Travel Writers Association publishes a "code of conduct" for its members on media tours. One tip: Avoid displaying bare midriffs in churches, mosques, or synagogues.[100] The website travelwriters.com exists as a sort of matching service for travel writers seeking tours and PR firms seeking travel writers.[101] The Corinth media tour is typical of the practice.

Corinth, population 14,000, is at the crossroads of two mainline railroads, and therefore Corinth was crucial to supplying Confederate armies during the Civil War. One of the fiercest battles of the war, Shiloh, was fought 22 miles northeast of Corinth, and a later battle, the Siege of Corinth, is still taught as the finest example of troop entrenchment. Some of the trenches remain, much as they were 150 years ago. Both Confederate and Union troops are buried in a national cemetery in Corinth. Civil War–era locales dominate among the 20 places listed on the National Register of Historic Places.[102] The National Park Service operates a Civil War Interpretive Museum in Corinth.[103] Tourism is important to Corinth: Visitors spend $42 million annually in Corinth, and tourism provides one in every 20 jobs and nearly $4 million in tax revenue.[104] It's not like the gaming halls in Biloxi, but Corinth is far smaller. So, as the war's sesquicentennial approached, tourism officials knew this was Corinth's time. But Corinth is remote. It's in Mississippi's hill region, far in the northeast. No scheduled airline serves Corinth. The Greyhound bus doesn't come through, and the nearest interstate highways are 42 miles away. Time to bring in the travel writers.

The Corinth media tour started with Kristy White, executive director of the Corinth convention and visitors bureau (CVB). White is a history graduate and former local history museum director. She took over as head of Corinth's CVB in 2006. White became a believer in media tours after the success of the first tour she sponsored with National Park Service money in 2007.

It was "huge," she said. The tour produced "$200,000 in editorial for $30,000 in cost," she said, applying a formula that equates a page of editorial with the cost of a page of advertising.

That formula came from Georgia Turner, whose travel PR firm organized the media tour for White. Turner's a media tour veteran: By the end of 2010, as Turner was winding down the business, she had led 1,200 travel writers on

160 media tours.[105] The Society of American Travel Writers honored Turner three times between 2004 and 2008 for service to the organization.[106]

Turner recruited eight travel writers for the Corinth tour. They came from Florida, Georgia, Washington, Texas, Pennsylvania, Wisconsin, and Massachusetts. One owns a lifestyle magazine, three are freelancers, one is a columnist, and three are editors. What stood out for Turner was that six of the eight had assignments. Often with tours, Turner said, the writers "might not write for years."

Not this group. Over parts of three days, they visited the Civil War battlefields and museums, historic buildings downtown, retailers, the crossroads of the mainline railroads, and a few local restaurants. At Abe's Grill, a sign on the door read, "Welcome Press Tour."[107] Like their colleague Karon Warren, they wrote about what they were shown, and they liked what they saw. Take, for example, the stop at Borroum's Drug Store. "By itself," one of the writers wrote, "it's a great reason to visit Corinth."[108] Another declared the milk shakes to be "the best . . . in the country."[109] "Two thumbs up from Wisconsin, y'all," wrote a third.[110] It is also true that the only critical note I found in the Corinth coverage applied to Borroum's: It was from Debi Lander, who tried the "slugburger" at Borroum's and "can't really recommend it." But, she wrote, "I can wholeheartedly rave about the milk shakes."[111] Lander, to date, is the most prolific of the writers on the tour—four published pieces from the tour with two still in the works. And she is the only one to disclose that she accepts "free products, services, travel, event tickets, and other forms of compensation from companies and organizations."[112]

The others don't disclose free travel. "I doubt my readers know what a media tour is," wrote Renee S. Gordon, "and I'm sure they are not interested." Besides, Gordon wrote, "I do not 'evaluate' or rate sites and attractions. I only list options."[113] In fact, Gordon's two-part article for the *Philadelphia Sun* evaluated 14 attractions in Corinth with 17 observations such as these: "It offers much more than . . . to a layman's delight . . . a good place to purchase . . . a one-of-a-kind treat . . . a must stop . . . renowned for its . . . a special aura of authenticity."[114] Another writer didn't disclose the free travel because it "has no bearing on [the] reader," Don Woodland wrote. "We're pretty honest about coverage." Sure, but consider what he wrote about chain hotels in Corinth: "A number of chain hotels are represented in the Corinth area. One of the best is the Holiday Inn Express. The accommodations are immaculately clean, roomy and the beds are incredibly comfortable." Do you think Woodland's readers would want to know that he and the other writers stayed free in the Holiday Inn Express and did not, as a group, inspect the other chain hotels?[115]

If the Corinth tour is typical of current media tour practice, it may also portend its future.

White said the next media tour to Corinth may target individual travel writers, not a group. She's thinking of working with Georgia Turner's former partner, Laurie Rowe, whose firm says it is "redefining" the media tour.

"You'll have access to a seasoned writer who is ready to generate publicity on your behalf," Rowe's firm promises destination marketers like Kristy White. Rowe guarantees "customized editorial content that encapsulates your destination" that will be published because of "industry contacts made possible only by years of experience on the editorial side."[116]

White thinks the new approach will work.

"I'm more likely to pay attention to an article written by a travel writer than an ad," White said. "I trust it a little more."[117]

## · 5 ·

# HOW IS GETTING "PUBLISHED" NO LONGER JUST ABOUT QUERY LETTERS TO NEWSPAPER AND MAGAZINE EDITORS?

Bob Jenkins got to travel journalism the hard way—from editing hard news.

Jenkins joined the *St. Petersburg Times* in 1969. He was editor of the National section during Vietnam and Watergate. He was a state news editor, a city editor. Then the paper decided it needed its first travel editor—and picked Jenkins.

Jenkins had a staff, a budget, and a weekly section to fill. His newspaper valued independence, ethical practice, and substantive coverage.

By all accounts, Jenkins was a superb travel editor. He and his Travel section won nine Lowell Thomas Awards, often referred to as travel journalism's Pulitzer Prizes. In 2006 the Society of American Travel Writers awarded Jenkins its highest honor, Marco Polo status.

But then the financial pressures bedeviling the newspaper industry took hold at the *Times*. Buyouts were offered throughout the newsroom. After 39 years—19 as travel editor— Jenkins was one of the ones to go.[1]

Today, Jenkins is on his own, an aging entrepreneur, traveling on subsidized press trips, writing for others and for his blog, bobjenkinswrites.com. As such, Jenkins is an exemplar. As traditional outlets for travel journalism dry up, travel journalists must go elsewhere. Getting published is no longer about carefully crafted query letters to guidebook, newspaper, and magazine editors. Increasingly it's about publishing online, either at someone else's site or your own.

"If you still think in terms of 'articles' and 'books' and 'pitches to editors' and 'publications,' you're stuck in the 20th century," writes online guidebook entrepreneur Tom Brosnahan. "And it's going to cost you . . . big . . . if you don't adapt."[2]

Travel journalists like Bob Jenkins aren't stuck in the 20th century. They're adapting.

## Getting Travel Journalism Published in Print

Much is known about getting travel journalism published in print. But print outlets for travel journalism are in steep decline—and are expected to decline further. Less is known about getting travel journalism published online, yet, online publishing is the fasting-growing outlet for travel journalism.

One purpose of this chapter is to resolve this conundrum.

Three good sources of information about getting travel journalism published in print are the dominant "how-to" books, submission guidelines, and survey data from travel editors. Taken together, the first thing to know is that getting travel journalism published in print—especially in the top guidebooks, travel-related magazines, and newspaper travel sections—is incredibly competitive.

Consider, for example, *Via*, the American Automobile Association's magazine for 4.5 million members in eight western U.S. states. *Via* was named a "best travel magazine" twice in the last decade by the Lowell Thomas Awards, ranking fifth among travel magazines in that competition. Chances are one in 50 of getting a feature story accepted by *Via*.

"We receive between 1,200 and 1,500 freelance queries and manuscripts each year," *Via* tells freelance writers in its submission guidelines. "We buy less than 2 percent of these."[3]

All of the top print outlets issue submission guidelines. Most, but not all, are available online. Read closely to understand what the outlet wants—and what it doesn't want.

Some typical issues:

- Don't pitch full-length feature stories unless you are an established writer already known to the outlet. Instead, offer shorter, Service and Advice stories for "the front sections." At *Via*, for example, this includes a where, what, how feature called "On the Road," an interview with an intriguing personality called "Trailblazer," or a "portrait of a visit-worthy section of a town or city" called "Neighborhood."[4]

- Policies on subsidized travel. Most of the top print outlets do not accept proposals or manuscripts based on subsidized travel. The language in the policies differs, but the point is the same. "We do not accept proposals about trips that are subsidized in any way," writes *National Geographic Traveler*.[5] At *Travel + Leisure*, "Neither editors nor contributors may accept free travel."[6] Same for *Budget Travel*,[7] *Caribbean Travel + Leisure*,[8] and *Conde Nast Traveller*.[9]

- What the reader wants—and doesn't want. All of the top print outlets commission third-party readership studies. They want to know what readers want and don't want. "*Afar*'s primary readers are active citizens of the world, curious and engaged," the magazine writes in its submission guidelines.[10] These are "college-educated men and women, ages 35 to 55, who have good jobs and high incomes." And what do they want? "They want to get beyond the superficial," *Afar* writes, "the mass-produced, the mass-consumed and the mass-experienced." Often, the guidelines cite wanted niches: "food, lodgings, ecotourism, adventurous learning experiences and short getaways," in the case of *National Geographic Traveler*.[11] Or, "no golf," in the case of *Via*.[12]

- Note the rules on "etiquette," the term of art pertaining to how these outlets want to be contacted. Mail and email are typically fine. Phone calls aren't. Some will respond to unsolicited queries and manuscripts. Others—famously, the *New York Times* Travel section[13]—won't.

Pitches take the form of query letters. *The Travel Writer's Handbook*, now in its 6th edition, devotes 28 pages of instruction to writing query letters to newspaper and magazine travel editors.[14] "The query letter is your foot in the door," another "how-to" book advises, "your chance to impress the editor with your perceptiveness and your prose."[15]

But what does the editor want? I asked editor members of the Society of American Travel Writers (SATW) to "rank the relative importance of [nine] qualifications of freelance writers who seek to publish work with you."[16] Their rankings, based on those who said the qualification was "somewhat" or "extremely" important, are as follows:

#1—"A feel for the character, content of your coverage"
#2—"Previous work for you"

#3—"The quality of the query"

#4—A tie between "Clips that demonstrate abilities you require" and "Still photography skills"

#6—"Does not have a financial or familial relationship with the subject"

#7— "Audio slide show skills"

#8—"SATW active member"

#9—"Video skills"

Several things stand out in the data. First, the data is consistent with advice in the submission guidelines of the top print outlets: know the outlet, have a relationship with the editors, write good query letters. Second, the data indicates that multimedia skills are still emerging as important qualifications and lag behind still photography skills. Third, not having a financial or familial tie to the subject is important. In fact, this qualification ranked second in the "extremely important" box. And, fourth, SATW membership is relatively unimportant to SATW editors.

## Getting Travel Journalism Published Online

It's no surprise the number of pages the "how-to" books devote to getting published in print. The surprise is how few pages are devoted to getting published online—just two in *The Travel Writer's Handbook*, for example.[17] This is remarkable for two reasons—first, because the outlets for travel journalism in print are in steep decline.

Take newspaper travel sections, for example. Circulation declined 35 percent between 2000 and 2010 for the newspapers whose Sunday Travel sections were the most honored by the Lowell Thomas Awards.[18] During the same period, circulation was mixed for the most honored travel-related magazines.[19] Circulation at the most honored magazine, *National Geographic Traveler*, was about the same, but its sister magazine, *National Geographic Adventurer*, went out of business. Circulation doubled at the next most honored magazine, *Budget Travel*, but was down or up modestly at four others.[20]

Top newspaper Travel section editors joined newsroom colleagues whose jobs were eliminated in the last decade. The *Seattle Times*'s John D. MacDonald left for a public relations job in government.[21] Robert N. Jenkins, Jr. left the *St. Petersburg Times* and became a travel blogger.[22] Jane Wooldridge of the *Miami Herald*, SATW's Travel Journalist of the Year for 2005–2006, stayed with the newspaper but moved to the business desk.

Worse still, but perhaps not a surprise, SATW members—both freelancers and editors—expect travel and tourism coverage in print will significantly decline over the next two to three years. Some 61 percent of freelancers who responded to my survey "somewhat" or "strongly" agreed with this outlook, as did 42 percent of editors who responded.[23]

The second reason it is surprising to see so little attention to getting travel journalism published online is because online is the fastest-growing outlet for everything travel-related. The audience for travel journalism online is large: The top 10 travel-related sites often attract 165 million visitors a month, according to Nielsen data.[24] Travel-related advertisers chasing this audience devote "a significant portion of their budget[s] to online advertising," also according to Nielsen data.[25] And their investment is paying off: Travel is the #1 category of online transactions.[26] Some 39 percent of all online transactions are travel-related.[27] Indeed, more travel is purchased online than offline.[28]

Yet, travel journalism was slow to adapt—and still is.

Back in 1997 the *American Journalism Review*'s new-media columnist, J. D. Lasica, took travel journalists to task for missing reader interest in travel online. "When it comes to newspaper websites," Lasica wrote, "the travel section is just an afterthought—if it's given any thought at all."[29] Back then, most newspaper Travel sections didn't post their content online. When they did, 100-inch features were shoveled onto sites, often without photographs— "flying in the face of the Web's very raison d'etre," wrote Lasica.[30]

It is some better today, but newspaper Travel sections and travel-related magazines still lag behind their competitors online. Only two newspaper websites—the *Boston Globe*'s boston.com and the *Anchorage Daily News*'s alaska.com—are among the 13 travel-related websites honored with a Lowell Thomas Award between 2000 and 2010. Just three of the 13 are travel-related magazine websites. Not one Webby Award, in any travel-related category over the last 14 years went to a site run by a newspaper or magazine.

So where does that leave the travel journalist hoping to get published online? Two options. One option is to get published by online travel sites owned by others. The other option is to develop your own site and publish there. Travel journalists are pursuing both options. The first is discussed below, the second in Chapter 6.

Here's a list of the 13 travel-related websites honored with Lowell Thomas Awards between 2000 and 2010. They are listed in rank order, beginning with the most honored:

- lonelyplanet.com, five awards
- nationalgeographic.com/traveler, four

- worldhum.com, three
- newyorkology.com, three
- budgettravel.com, three
- cruisecritic.com, two
- boston.com/travel, two
- frommers.com, two
- reidguides.com, one
- turkeytravelplanner.com, one
- wildwritingwomen.com, one
- alaska.com/akcom/travel, one
- southernliving.com/southernbyways, one
- matadornetwork.com, one

This is a strong list, a good cross-section of consumer-facing, content-oriented sites. But do they offer good opportunities for travel journalists? Look more closely. Some of the sites, such as expedia.com, are transactional sites and not outlets for travel journalism. Others such as turkeytravelplanner.com and reidguides.com, are single-authored, entrepreneurial sites not open to other travel journalists. Same for wildwritingwomen.com, open only to its 12 authors. Worldhum.com, perhaps the most celebrated for its long-form, literary approach to travel journalism, is no longer accepting submissions. Lonelyplanet.com, the most honored site, is taking applications from freelance guidebook authors, but accepts fewer than 2 percent, about the same acceptance rate as magazine-related sites such as budgettravel.com and nationalgeographic.com/traveler.

And then there is matadornetwork.com. Honored by the Lowell Thomas Awards for the first time in 2010, Matador is the most accepting of freelance offers, but pays almost nothing. Yet, websites such as Matador offer travel journalists the best opportunity to get published online by sites owned by others. So it's important to understand how Matador and sites like it work.

Matador bills itself as both the "best online travel magazine" and the "best online travel community."[31] Fulfilling this double-barreled promise requires that Matador operate a continuous cycle of training, work assignments, and publication. It's a model for travel journalism that "other fields of journalism should examine," according to faculty at UNC Chapel Hill who wrote Matador's Lowell Thomas Award citation.[32]

Here's how it works:

The cycle begins with training. Matador and its affiliate sites recruit would-be travel journalists for training at Matador U, an online training center offering courses for writers, photographers, and filmmakers.

Reporting, broadly understood, comes next. Matador matches graduates with opportunities to travel and report. For some, it's grants, fellowships, and residencies. For others, it's PR firms and tourist boards for expense-paid press trips. For some, it's assignments from Matador editors or editors of other travel publications. And for still others, it's tools to create travel blogs of their own.

The cycle ends with publishing. Matador, alone, has 11 outlets. They range from destination guides to service and advice articles to journey articles. Other travel publications appear eager to work with Matador graduates. *National Geographic Traveler*'s editor Keith Bellows, for example, says he "is looking to Matador to help us find new talent."[33]

Along the way, Matador reinforces the idea of community. Once enrolled, students have "lifetime access" to courses, workshops, forums, and blogs. Matador wants to link travelers with nonprofits: The site profiles 68 pages of opportunities. Matador also wants to link travelers to each other, with some 3,000 pages of "active members" who "friend" and message one another.[34]

How well is it working? Alexa.com traffic rankings confirm that the site enjoys a wide reach—its global traffic rank is 7,188.[35] To see that in context, top-rated lonelyplanet.com ranks nearly five times higher.[36] But the highly rated budgettravel.com ranks four times lower.[37] Matador's audience is attractive to advertisers: It is younger than the general Internet population, female, well-educated, and childless.[38] Search data indicates the audience is entrepreneurial: The most popular search criteria used to reach the site was "make money traveling."[39] At $25 per 2,000-word article, they won't make much money at Matador, but it's a starting point.

They won't make much money at the other Matador-like sites, either. That's partly why Sarah Stuteville discourages most would-be travel journalists from trying. In her essay "To Be (or Not To Be) a Travel Journalist," Stuteville recounts her first year as an independent journalist, publishing online at sites such as the Common Language Project, which pays $50 to $100 for stories, depending on medium, length, and complexity.[40] "After my first year," Stuteville says, "I feel it would be unfair not to start by trying to dissuade you."[41]

But what is the next level? Is there any reason to believe that sites such as Matador will prosper to the point that it will offer $1 per word or more, as all of the top-rated travel magazines offer, instead of the penny per word Matador offers today? That's 10 orders of magnitude, a steep curve. Is the next level a transition to more traditional publishers such as *National Geographic Traveler*, whose editor says he's looking to Matador for talent? Or is the next level something not yet developed?

## Test Your Understanding

1. Identify the four submission guidelines typical of print outlets.
2. Look again at the data about freelancer qualifications most in demand among editors. Which of the nine align well with your qualifications? Which do not?
3. Experts say travel journalism was slow to adapt to online opportunities. Why do they say this?
4. Describe the three-step process Matador Network follows in publishing the work of freelance travel journalists.

## Practice Your Skills

Assignment #5: Select an outlet, either print or online, for the story you outlined in Chapter 4. Draft a query letter, as appropriate, based on the outlet's submission guidelines. Consider the most-in-demand qualifications discussed in this chapter. Assess how your experience matches up. Incorporate, as appropriate, in the query letter. Ask a fellow student, teacher, colleague, or editor to evaluate it and provide feedback.

---

## A Closer Look: How You Might Construct a Travel-related Website

Imagine wanting to build a travel-related website associated with towns in the U.S. dominated by the military—so-called "military bastions"—a personal interest. Imagine, too, that you're new to travel-related content, site development, and military bastions. But you are entrepreneurial and eager to begin.

The purpose of this case study is to provide a model of how you might do this.

Let's begin with some definitions:

Travel-related content—Let's use the four travel journalism story types identified in Chapter 3. They are: News, Service and Advice, Destination, and Journey. Any one or all are fair game for this case study. In terms of a topic niche, think military travel.

Site development—We'll understand *site* to mean anything from the simplest blog to the most complex network of full-featured websites. *Development* means building a site on your own or with someone else's help.

Military bastions—The term is from Patchwork Nation, a multi-organization journalism project led by the Jefferson Institute's Dante Chinni. The project assigns one of 12 "community types" to the nation's 3,143 counties, based on demographic characteristics. One type is "Military Bastions," defined as "areas with high employment in the military or related to the presence of the military and large veteran populations."[42] Just 55 counties across 23 states qualify.

Given these definitions, the first step blends curiosity with research. From a travel-related content perspective, what piques your curiosity about military bastions? Ask three friends to join you for a brainstorming session. When I did this, these questions emerged:

- Relative to other "community types," do military bastions attract above average, below average, or an average number of tourists?
- What can we learn about the audience for military bastion travel-related content?
- Who serves this audience today, in print and online?
- Can we identify an underserved interest?
- Can we identify travel-related content that would meet this under-served interest—online?

These are questions that are typical of the idea-development process. We don't, yet, have an idea. What we have are questions that might lead us to an idea. We know from innovation best practices discussed in Chapter 6 that research is the next step. What sources would we use to answer one or more of the five questions?

- Start, perhaps, with the convention and visitor bureaus (CVBs) in the 55 "military bastions." Email a simple questionnaire using a low-cost, online survey tool such as Zoomerang or Survey Monkey. They should have good data and insight on all five questions.
- Demographic and psychographic data is available from three sources. First is the Patchwork Nation website, patchworknation.org. Second is the county-level data at the Census Bureau's American Community Survey. Third is the SRDS *Lifestyle Market Analyst* introduced in Chapter 2.
- Who serves this audience today, in print and online? Start with local news media available online. Each is likely to have a reporter,

editor, or producer assigned to the "military" beat. Search sites under this person's byline. Scan the coverage. Consider emailing a simple questionnaire as with the CVBs. Large military installations publish weekly newspapers available to the public in print and online. Scan the coverage of travel-related topics: graduations, reunions, museums, cemeteries, competitions, memorials, demonstrations, and so forth. Each of the services is covered by a national trade publication such as *Army Times*. Search military-related websites on a ranking site such as alexa.com. Number #1 is military. com. Scan its travel- and tourism-related content.

What underserved, travel-related interest might emerge from the research? For purposes of the exercise, let's just pick one: military-related museums. Deep within the pages of military.com is a state-by-state list of military museums. A quick scan shows at least one museum in each of the 55 military bastions. But there's no content on the military.com site, just links to the museum's websites, a mishmash. A Google search indicates that no other site aggregates content about military-related museums, either.

An underserved interest? If so, the questions to ask now are these:

- What travel-related content might satisfy this interest? Information about the permanent collection as well as temporary exhibitions? TripAdvisor-like reviews from patrons? An online link to the museum's gift shop? Recommendations about travel, accommodation, food?
- Might the start-up focus on one museum as a way of testing, proving up, the concept? What's the platform for the start-up? A simple blog with user comments? Applications to engage the user community, such as Facebook and Twitter? A simple way to manage relationships with users, such as Constant Contact?
- Can you afford to self-fund the start-up? If not, are there museum-related foundations that might provide a seed grant? Military-related foundations? How about a Kickstarter project?
- Going forward, how will the site make enough money to sustain itself—and you? Grants and Kickstarter-like funding are fine for start-ups, but not for ongoing support. For that, you'll need advertising, user payments, or a marketing alliance with the gift shop.

- A value proposition should emerge from the start-up, one that is strong enough to justify scaling up the site to include some, most, or all of the military-related museums in the 55 military bastions.

Keep in mind the lessons from innovation: Most fail. Risk increases with time. So fail fast. Remember the three reasons why most innovations fail: Insufficient or faculty audience research; poor test marketing; inadequate financial analysis.

Ready, aim, fire.

# · 6 ·

# WHAT ARE THE OPPORTUNITIES FOR THE ENTREPRENEURIAL TRAVEL JOURNALIST?

Meet Matt Kepnes, better known to travelers by his travel blog's URL, nomadicmatt.com. Kepnes is the latest among a string of travel journalists who've succeeded as entrepreneurs.

Success for Kepnes and entrepreneurial forebearers such as Lonely Planet's Maureen and Tony Wheeler, Matador's Ross Borden and Ben Polansky, and WorldHum's Michael Yessis and Jim Benning means several things. First they are living their lives in travel, independent of someone else's business and their rules. Second, they are earning a good living doing so. And, third, they are encouraging others to do the same.

Journalism wasn't Kepnes's starting point. That was hospital administration.

Kepnes worked a cubicle job in Boston, his home, and was pursuing an MBA when a career-break trip to Thailand in 2005 changed his course. "My original trip was supposed to last a year. I didn't come home until 18 months later," Kepnes writes in his site's "About Me" page. "Once back, I knew I couldn't go back to my old life or a typical job—I wanted to travel. I wanted to make this my life. Three months later, I was on the road again and I haven't stopped since."[1]

By early 2008 Kepnes had created his core travel blog, nomadicmatt.com. A set of travel-related advice e-books came next. By late 2009 revenue from these ventures was $7,500 to $8,000 a month, Kepnes told *New York Times Frugal Traveler* columnist Matt Gross.[2] In 2011 the site attracted 120,000 unique visitors a month, earning an alexa.com Traffic Rank of 26,914.[3]

Could other travel journalists do as well? Kepnes—and many others like him—think so. "If I, a lazy guy from Boston, can make this travel thing be affordable and work for me," Kepnes writes, "you can do it too!"[4]

What they need to learn is the point of this chapter: Opportunities are varied, significant, and enduring for the entrepreneurial travel journalist. The approaches are well understood. Examples of best practices abound. Advice is excellent and accessible. The key is having the right mindset, learning the steps, and executing them. The caution is understanding—and avoiding—the legal and ethical pitfalls.

Mindset is important—maybe the most important. There are two parts. One part is having the mindset of an innovator. The other part is having the mindset of an entrepreneur. Neither mindset is typical of journalists, but many journalists—including travel journalists—are learning.

In this chapter, you'll learn from them.

## What We Mean by the Mindset of an Innovator

The innovation mindset is little more than the "capacity to conceive, develop, roll out and improve new offerings."[5] Experts typically distinguish between two types of innovation. One type is "sustaining innovations," which bring better products or services to established markets and customers. The other type is "disruptive innovations," which transform the demands of established markets and customers and, in so doing, disrupt current providers.

To one degree or another, all businesses innovate. Most innovations are of the first type, sustaining innovations, often a matter of an existing business getting better at what it's already doing. The second type is rarer—and scarier—because disrupting innovations are what are done to existing businesses by others. Mapquest's online directions app, for example, disrupted AAA's TripTik Travel Planner business. Why troop off to the local AAA office, then wait a few days for a clerk to assemble a spiral-bound TripTik, when Mapquest provided a comparable product in a matter of moments, online—for free?

Where do innovative ideas come from? "New to the world" ideas are rare. Indeed, air conditioning, the personal computer, and disposable diapers are remarkable innovations because of their rarity. More common are ideas that extend, repurpose, or replicate existing ideas. (Travelocity, Expedia, and Orbitz disrupted travel agents and airlines that controlled air travel reservations, ticketing, seat selection, and boarding passes. But the underlying technology—for example, the Sabre system—already existed.)

Executing good ideas is difficult, but well understood:[6]

- Locate your idea on the "innovation continuum." The continuum ranges from "low innovations" such as improving processes to reduce costs and repositioning a product for a new market, to "high innovations" such as product lines that are new to a company or products that are new to the world. "Low innovations" involve the least risk but the lowest return. "High innovations," by contrast, involve the most risk but the highest return.

- Recognize that most innovations fail. On average, across all industries, for every 11 serious ideas or concepts, three enter development, 1.3 are launched, and one succeeds. Risk increases with time, as costs rise rapidly as an idea progresses through development, testing, and launch. Success rates vary based on newness. "Low innovations" tend to be familiar, not new, and have high success rates. "High innovations" are just the opposite. So, the mantra among innovators is to "fail fast," before costs mount in pursuit of a bad idea.

- Avoid the common causes of innovation failure. Three reasons stand out in 50–75 percent of failed innovations: insufficient or faulty marketing research, no test marketing (or poor test marketing), and lack of adequate detailed financial analysis. "Every few years, innovation resurfaces as a prime focus of growth strategies," wrote Harvard's Rosabeth Moss Kanter. "And when it does, companies repeat the mistakes they made the last time."[7] Learn and avoid.

As a test of your innovator mindset, ask three questions of yourself. First, what is your readiness to innovate? The Readership Institute at Northwestern's Medill School of Journalism developed a "Ready to Innovate Index" that measures readership mission, customer service orientation, responsiveness, and adaptability.[8] A similar, but simpler, self-test was developed by Clayton Christensen's consulting firm, Innosight, for the American Press Institute's Newspaper Next project. Sample agree-disagree question: "We can accept and run with ideas that were 'not invented here.'" Answers to the 20 questions provide a readiness-to-innovate score.[9] Second, what is your appetite for failure? Failure, of course, isn't the goal. But innovators must be able to tolerate failures.

"Embrace experimentation, not failure," urges IDEO's and Stanford's David Kelley. But "tolerate small failures to win big."[10] And, third, are you persistent? Innovation requires a lot of trial. But trial—or trying—isn't enough. It's succeeding— or doing—that matters. I like what Navy SEALs say about trying and doing: "There is no try. There is only do. Do or do not. There is no try."[11] What a mindset.

## What We Mean by the Mindset of an Entrepreneur

Most journalists are not entrepreneurs. Rather, they are employees of companies where others—those people on the "business side"—take responsibility for profit and loss. Few travel journalists are employees. Travel editors are typically employees, as are their copy editors, but not the writers and photographers. Sixty percent of the travel editors who responded to my survey reported employing no full-time travel writers. Another 18 percent reported employing one.[12]

Most travel journalists *are* entrepreneurs, or what I call "almost entrepreneurs." They are freelance writers and photographers who seek assignments from or sell work to newspapers, magazines, and guidebooks, either in print or online. While they are not employees drawing a paycheck, neither are they independent travel-related businesses, responsible for profit and loss. More accurately, to use business school language, freelancers are part of the travel industry supply chain, providing a part in the process. They are entrepreneurs in the sense that they assume the risk for their personal success or failure, but they are not the creators, the independent owners of the enterprise—hence, "almost entrepreneurs."

An increasing number of travel journalists are true entrepreneurs, owners building enterprises. Some 42 percent of the freelance travel journalists who responded to my survey reported that they publish travel-related blogs and websites.[13] An examination of their sites shows a broad range of enterprise with respect to content, revenue, and networking. Some are little more than story archives. Others are robust, full-feature sites with outside contributors, multiple revenue streams, and active, frequent use of engagement tools.

Some details from the survey:

- While 57 percent are simple blogs, the rest are web-like sites on blog platforms or websites. While 84 percent publish work by a single contributor, 16 percent have multiple contributors—including "guest bloggers."

- At 68 percent, Service and Advice and Destination type stories dominate content. News accounts for 29 percent of content, and Journey accounts for 3 percent.
- Some 58 percent attract some form of revenue. Advertising dominates at 59 percent, followed by sales at 35 percent and fundraising at 5 percent. Only one site attracted revenue from all three. Some 42 percent have no source of revenue.
- Some 69 percent of the sites engage users through subscriptions and syndication. Some 31 percent use one or more forms of "sharing," and 20 percent use comments. Some 49 percent engage users through Facebook or Twitter. While 20 percent engage users through both, 71 percent use neither.

Clearly, some of these travel journalist entrepreneurs are farther along than others. That should come as no surprise. Journalists are unaccustomed to entrepreneurship and its low-cost, garage-like tinkering.[14] Journalist and "change advocate" Michele McLellan has identified three disadvantages that hinder journalists in becoming entrepreneurs: They are traditional thinkers who prefer "perfecting the familiar" to invention. They are "squeamish about business" and its imperative to produce a good or service that will sell. And, they are "math-phobic," a fatal condition for an entrepreneur.[15]

But McLellan and many others reject the idea that these disadvantages are disqualifying. McLellan, for example, cites brains, persistence, and networks as advantages for journalists wanting to be entrepreneurs.[16] J-Lab founder Jan Schaffer says comfort with change, a curiosity about technology, and an understanding of what makes a business viable are essential qualities.[17] "Savvy and hustle are the most important," says David Cohn.[18] But you can learn, as many journalists are learning. Whether it's from books, classes, boot camps, or graduate-school programs, many are.[19]

Side by side with entrepreneurship is a different ethic, a different standard for journalism. Traditional journalism is about authority and control. "A significant portion of serious people," writes Bill Keller, "feel the need for someone with training, experience and standards—reporters and editors—to help them dig up and sort through the news, identify what's important and make sense of it."[20] Quoting *Guardian* editor Alan Rusbridger, Keller writes that authority and control are in tension with "a world in which many (but not all) readers want to have the ability to make their own judgments, express their own priorities, create their own content, articulate their own views, learn from peers as much as from traditional sources of authority."[21] The latter view tracks with the Google News approach, which explicitly rejects the value of thousands of outlets covering news in the same way. "Usually, you see essentially the same approach taken by a thousand publications at the same times," says Google News founder Krishna Bharat. "Once something has been observed, nearly everyone says approximately the same thing."[22] Traditional journalism is less and less able to underwrite such inefficiency. Entrepreneurial journalism won't tolerate it. Thus, new standards emerge. Objectivity is the standard for traditional journalists; transparency is the standard for entrepreneurial journalists. Traditional journalists are aloof; entrepreneurial journalists are engaged. Traditional journalists attribute; entrepreneurial journalists link. "Make us proud," say traditional journalists. "Make us money," reply entrepreneurial journalists.[23]

## The Travel Journalist as Innovator and Entrepreneur

We know from the survey data cited above that 42 percent of freelance travel journalists who responded publish travel-related blogs and websites. And from studying their blogs and websites, we know something about the range of enterprise with respect to content, revenue, and engagement.

But there are additional sources of evidence of the travel journalist as innovator and entrepreneur, the approaches they are taking, the content they are exploring, the results they are achieving.

How might you use this information to find a place to start?

Approaches are based on cost and knowledge, and fall within three categories:

- What you can do for free, on your own. Begin with a simple blog, published on a free platform such as Blogger or WordPress. Create some of the content, aggregate the rest. Add Flickr for photos, YouTube for video—both are free. Add Google's AdSense for revenue—though you won't have much until you develop some traffic. Add Facebook and Twitter, and respond to comments to drive traffic and to increase engagement.
- What you can do for little expense, with a little help from others. Simple blogs are just that, simple. Yours is the same as the next travel blogger who chose the same theme. You need not know how to write code to achieve any of the outcomes listed above. No HTML, no CSS,

no XML. Each application is a mouse click or a drag-and-drop away. But you do need to learn how to write code—or hire someone who does—in order to exploit the potential of a Blogger or WordPress blog. Instead of sending users to Flickr or YouTube, bring photos and video to your site. Instead of relying on AdSense to serve Google's advertising messages, sell and serve your own.

- More sophisticated applications come with higher cost or knowledge. A window of travel-related news feeds. A store that sells travel-related products. An offering of travel-niche topic blogs, created by others for your site. Deals of the day tied to location and loyalty. A reservations app.

Within each approach there are thousands of examples to study. Alexa.com lists 6,411 "top sites" in 13 travel-related categories. Two percent of the million-plus blogs listed on Technorati relate to "tourism," tens of thousands more to "travel." Maybe it makes sense, among all these numbers, to look at who seems best-at-what in various categories—content-oriented blogs and websites, mobile applications, use of social media to drive traffic, and engagement. Consult the top-rated sites. Use them as thought starters for your own work. What might you extend? What might you repurpose? What might you replicate?

## A Sampling of Travel-related Enterprises

Here is a sampling of travel-related enterprises based on 12 different approaches. Ask yourself this: Which ones come closest to your interests, skills, and experience?

## Constructing a Travel-related Enterprise Based on "Network Effects"

OpenTable.com is an example of constructing a travel-related enterprise based on "network effects." The idea draws on Metcalfe's Law, which holds that the value of a network equals the square of its nodes. If two people use OpenTable to reserve a table for dinner, the network is worth four; if four people use OpenTable, it's worth sixteen.[24] Like every network effects enterprise, OpenTable gets better as more people use it. Free to diners, OpenTable makes money by charging restaurants for tables booked—some $100 million in 2010.[25] "By aggregating the world's restaurant goers," writes Fast Company's Farhad Manjoo, OpenTable "has changed the nature of the relationship between eateries and their best customers by inserting itself in the middle."[26] Next step: Offer its large corps of loyal users other travel-related goods and services. A limo ride to the restaurant? Flowers on the table? Rewards for paying the dinner bill on a particular credit card? Oops! Already doing it.

## A Site for Literary-quality Travel Journalism

WorldHum.com is an example of constructing a travel-related enterprise that links literary-quality travel journalism and technology—with a big payoff at the end. Jim Benning and Michael Yessis started World Hum in 2001 with one goal: "To publish the best travel stories on the Internet." A decade later, it's widely regarded as one of the top outlets for literary-quality travel journalism—online or in print. The focus is on Journey-type stories. World Hum won a Lowell Thomas Silver Award for best travel-related website in its first year in business, and two more since. World Hum stories are regularly anthologized in *The Best American Travel Writing*. "Superb writing and stylish layout make visiting the site like cracking open a high-quality travel magazine," wrote the *Wall Street Journal* in 2006. In a blog post citing the article that day, Benning noted that the *Journal* recommended such niche websites to investors. "Not that we're suggesting any investors take an interest in World Hum or anything," Benning wrote. "That would be crazy."[27] Less than a year later, the Travel Channel acquired World Hum for a large, undisclosed, sum.[28] Benning and Yessis remain.

## Short-form Travel Journalism Taken Seriously

Microblogging is travel journalism's shortest form. Don't be surprised that journalists take the form seriously: Alberto Ibarguen, chief executive officer of the John S. and James L. Knight Foundation, is a member of the board (and helps fund) the group that honors microbloggers on Twitter in 31 categories, including travel. Other journalists on the board include David Pogue of the *New York Times* and Jeff Jarvis of the City University of New York. The awards are the Shorty Awards, now in its third year. The awards are valuable as one window on to how travel-related enterprises—journalistic and otherwise—use microblogging services.

The 2011 winner was J. D. Andrews, an entrepreneurial travel journalist who claims to have written from 76 countries over six continents.[29] On the June 2011 day I followed Andrews on Twitter, he tweeted 63 messages over 24 hours. "I tweet a lot," Andrews wrote in his 140-character profile for the Shorty Awards.[30] Andrews uses Twitter much as the rest of the world does: to report current status, links to other content, product recommendations, and to retweet messages from others. About one-third of the top 25 Shorty Awards in travel went to travel journalists such as Andrews. But about two-thirds, including four of the top 10, went to destination marketing organizations such as the official guide to Columbia, South Carolina, or "I Love NY." This is consistent with research showing that microblogging services are increasingly dominated by marketers.[31]

# Liveblogging: Covering Travel as It Happens

A criticism of travel writing is that it lacks one of journalism's essential elements: currency.[32] Kevin Kerrane and Ben Yagoda omitted travel writing from their anthology of literary journalism for just this reason. They insist that "a writer get on the story soon after it happens."[33] When the story emerges months after the travel, they contend, it is more in the category of history than journalism. An extreme antidote to this problem is "liveblogging," which is an up-to-the-minute account of an event. Liveblogging is a subset of "beatblogging," a term popularized by Jay Rosen, a journalism professor at New York University and author of PressThink.[34] According to Rosen, a beatblog is any blog that "sticks to a well-defined beat or coverage area"; is the work of "a single person or a team"; is authored by "a pro or an amateur journalist"; presents "a regular flow of reporting and commentary"; and provides "links and online resources" in order to "track the subject over time." At its best, a beatblog is a "two-way knowledge system that feeds the beat."[35]

Among beatbloggers, USA Today travel journalist Gene Sloan is a star. In 2009, for example, Sloan liveblogged while cruising on a refurbished Carnival Cruise Line Fantasy-class ship. As Mark Briggs describes in Journalism Next, he spent five days posting his impressions and answering reader questions.[36] "The ability for users to leave comments and suggestions," Patrick Thornton wrote on the best-practices section of beatblogging.org, "makes this form of journalism much more interactive and engaging for users."[37]

## Seeing the "Worst" in Travel

Seeing the "worst" in travel goes against the grain. But according to travel journalist Doug Lansky, it's an antidote to the PR-inspired, brochure-like content that comprises 80 percent of travel writing. Enter Lansky's Titanic Awards, dedicated to celebrating the "often spectacular underachievements in the travel industry."[38] Awards are made in seven categories, from worst accommodation to worst toilets. Nominations come from travelers who complete a survey at the Titanic Awards website and from travel journalists. This isn't an ombudsman site. Lansky doesn't propose to get any traveler's money back. It's an amusing, even playful process, shining a bright light on the "worst" of travel. But a serious purpose underlies the fun: holding the world's largest industry accountable. "There's just too much carrot in this industry and not enough stick," Lansky says. "This is a fun way to hold their feet to the fire."[39]

## Innovating the "Agony" Out of Travel-specific Search

Four of 10 online transactions are travel-related.[40] Almost all of them begin with a search, typically for flights, but also cruises, hotels, and rental cars. The earliest flight-search sites were innovators such as Travelocity, Orbitz, and Expedia, which permitted travelers to search flight data aggregated from the airlines, then book through the flight-search site. The idea was to show travelers choices—sometimes hundreds of pages of choices—based on cost, routing, and so forth. Travelers liked the choices, but loathed the slog. Then came the meta-sites such as Kayak, which continually scoured sites such as Travelocity, Orbitz, and Expedia, as well as the airlines, based on a single criteria important to most travelers: cost.

Now comes Hipmunk, seeking to simplify the process further. Hipmunk gathers the same data as the other travel-search sites, but screens it and presents it to travelers according to an "agony" algorithm that ranks flights according to price, duration, and stops. "With its clean interface and customer-first approach," writes Fast Company's Luke O'Brien, Hipmunk "is primed to zoom past all of them."[41] That is, until the next innovation arrives.

## Pro-am Travel Journalism

Like so many other genres today, travel journalism is not for professionals alone. Whatever term you apply—*citizen journalism* as an alternative to *pro-am journalism*, for example—the idea is that a team comprised of amateurs and professionals will report, produce, and publish content. As Dan Gillmor famously noted, we're all experts at something.[42] Pro-am journalism attempts to harness the amateur part of this equation.

CNN's iReports is an early and important example of the idea in play. Since 2006, amateur iReporters have submitted 604,000 iReports, of which CNN producers have "vetted" 46,500 to appear on one or more CNN platforms.[43] Travel is one of 12 iReport categories. CNN producers accept and consider what iReporters upload, but also put out assignments such as "travel photo of the day." The most recent travel-related innovation is "Mark your points of interest," a partnership with the location-based app Gowalla, which allows CNN to map where iReporters are traveling and where they are reporting about travel. No payment, but exposure on CNN.[44]

A more granular, geographic approach to pro-am journalism is examiner. com, part of the Anschutz family of companies.[45] It solicits local reporting in 25 categories from "examiners" across the U.S. Travel is one of the categories. Contributors earn a publishing outlet and a small split of the revenue. A subsidiary, NowPublic, offers the same opportunity for international coverage, including travel.[46]

## Daily Deals Shifting How Travelers Search for Bargains

All manner of travel deal sites populate the web. Travelzoo, for example, has 23 million subscribers taking advantage of its "recommended travel deals" from 2,000 companies.[47] Travelers are encouraged to sign up 10 days to several weeks before they travel in order to incorporate emailed offers into their travel plans.[48] Yelp and Yipit offer the same. Newer deal sites offering "daily deals" are shifting how travelers and others search for bargains. Some, such as Groupon and Living Social, depend on the traveler signing up for a location. Others, such as location-oriented Foursquare, feed deals to travelers wherever in the world they "check in" through their GPS-enabled smartphones.

Daily deal sites have "changed the business landscape," Nielsen estimates, "shifting how consumers search for bargains and connect with businesses large and small, national and local."[49]

## Using Geo Tools to Engage Virtual Travelers

Virtual travelers experience the world in every way—except by going there. Previously, they were the "armchair travelers" entertained by literary travel narratives from Evelyn Waugh, Graham Greene, and Norman Lewis. Today, they are increasingly engaged by online developers using the geo-tool products Google Earth and Google Maps. One is Google Sightseeing (unaffiliated with Google), which offers satellite and street views in 137 countries across the seven continents. The views are sorted by country or category; among the 28 categories is "abandoned." Here you'll see, for example, satellite views of wrecked planes at the airport in Luzamba, Angola, which Google Sightseeing calls "plane-wreck central." The site asks, "Why bother seeing the world for real?"[50] For travel journalists, the opportunity is to become a Google Sightseeing "author," offering ideas for as-yet-unseen venues.

## Travel Blog Aggregation: A Service to Travelers?

Aggregation, or "curation" as some bloggers prefer, is the practice of blogging about content created by others. It is a service consistent with what economists Carl Shapiro and Hal R. Varian say is an important purpose of journalism. "Nowadays the problem isn't information access but information overload," they write. "Real value . . . comes in locating, filtering, and communicating what is useful to the consumer."[51] Certainly, travel-related information is on overload.

Opportunities seem endless. One might aggregate by topic, by geography, or by audience segment. One might sort or filter by rating, by ranking, or by indexing information.

How is this journalistic? It is journalistic if it is timely, if it is credible, if it is independent, if it is substantive. It goes to a journalistic purpose of helping users make good decisions, including "wise purchasing of goods and services."[52] How to do this well? There are many step-by-step guides. Some, such as *Journalism Next*,[53] embrace language common to journalists. Others, such as *Curation Nation*,[54] are more general. Either will get you started.

Before you begin, be aware of the legal and ethical issues swirling around aggregation. Both *Journalism Next* and *Curation Nation* address them.

## Travel Video at the Intersection of Two Trends

Consider the intersection of two trends. One is the growth in the number of travelers who research and book travel online. More than 83 percent of U.S. travelers do so.[55] The other is the growth in the audience for online video. Three-quarters of U.S. online users watch more than two videos a day.[56] Not surprising then that *Gadling*'s Chris Owen concluded that "travel video is really popular."[57]

At the upper end is the Travel Channel, with so many professionally produced videos that it segments them four ways: featured, recently added, most liked, most viewed. A tier below are the news and information sites such as CNN that post professional and amateur-produced videos under a travel channel. Then there are the online travel magazines: Matador Network, which produces the largest online travel magazine based on audience, claims it "curates the best travel video on the web."[58] New, niche entrants are emerging. One of them, *GoingSocialTV*, merges travel video with social media. Much travel video is explicitly commercial, exploiting the fact that six of 10 users had taken some action after watching an online video advertisement.[59] A good example is *TVTrip*, a hotel video guide. Launched in mid-2007 by former Expedia executives, it boasts professionally produced video reviews of more than 6,000 hotels worldwide.[60] It's a promising approach, and five competitors using the same model have already emerged.[61]

But what of more journalistic fare? All of the highly awarded travel websites post videos, most of them in the form of destination guides. A clever, anecdotal offering is Robert Reid's weekly YouTube post, the *76-second Travel Show*. A typical episode: "The Great Airport Shoe-Removal Debate." Long-form travel-related documentaries are rare and difficult to find, in part because search tools for video files remain primitive.[62]

## Building a "Travel Community" Online

Are there places on the web where travelers come together as a community? Is building a travel community site an opportunity for an entrepreneurial travel journalist? The answer to both questions depends on a definition: What do we mean by "community" online?

Wikitravel, built on the wiki model of collaboration and consensus, would say it is a community. Begun in 2003 as an open-content travel guide, Wikitravel has grown to 50,000 articles and as many users across all versions. It is not just a community of travelers—business owners, local advocates, and tourism professionals are welcome to contribute, and do. Nor is it edited to a "neutral point of view," as is its model, Wikipedia. Rather, it is edited to a standard of "be fair." Wikitravel says its mission is to be "reliable" and "complete," but to "call a spade a spade." "If another Wikitraveller disagrees," Wikitravel says, "the description should be edited until both sides agree that the description is fair."[63] Wikitravel won the 2007 Webby Award as Best Travel Website,[64] and was named one of the 50 Best Websites of 2008 by *Time* magazine.[65]

A travel network such as Matador promotes itself as "a worldwide travel community for creating and sharing ground level media."[66] It lists 3,000 pages of member profiles with "friend" and "message" tools to promote interaction. Fifteen member forums cover such topics as finding travel partners and how to volunteer while traveling. Matador hosts member travel blogs and photo galleries. A "Bounty Board" lists paying assignments for travel writers—most of them on Matador-owned sites.[67]

Recommendation sites such as TripAdvisor and its rivals refer to their users as a community. TripAdvisor is the best known of the 18 travel brands operating under the TripAdvisor Media umbrella. Its 20 million members have posted more than 45 million reviews and opinions, making the company a world leader in user-generated content. TripAdvisor depends on travelers trusting and cooperating with one another, much as James Surowiecki described in his 2004 book *The Wisdom of Crowds*.[68] No central authority controls their behavior. Rather, "wisdom" emerges, in Surowiecki's terms, when TripAdvisor aggregates the informed, independent opinions of travelers. The Law of Large Numbers bears on this, as well. TripAdvisor, no doubt, feels comfortable declaring Prague's Golden Well Hotel #1 among the "Top 25 Hotels in the World" when 736 of 744 reviews are excellent or very good.[69] TripAdvisor employs moderators to examine reviews questioned by the owners of the businesses reviewed. Monitoring tools on the site help, too. But the traveler community matters the most: "Our large and passionate community of millions of travelers keep an eye out on our site as well," TripAdvisor says.[70]

Travelblogexchange.com, travelwritersnews.com, and travelwriters.com are examples of communities for travel journalists—and, in some cases, those who seek to influence travel journalists. Travelblogexchange.com is the work of Kim Mance, editor of *Galavanting*, an online travel magazine, and host of its web TV series. At its core is an annual meeting focused on how-to workshops. Some 450 attended the 2011 meeting in Vancouver. The meeting's organizing committee reflects the organization's stature. It includes such travel journalism luminaries as *Gadling*'s Don George, Lonely Planet's Robert Reid, World Hum's Jim Benning, and Vacation Gals's Kara Williams.[71] Travelwriternews. com is an example of a newsletter for a region of travel journalists—in this case, the San Francisco Bay Area. It is the labor of travel journalist Laurie McAndish King, who lists editorial opportunities, classes, readings, and news about Bay-Area travel journalists. "The feedback I've received," King says, "is that it's not only a useful calendar of upcoming events [but it's] building our growing SF Bay area community of travel writers and photographers."[72] Travelwriters.com is a matching service for travel writers and the editors and PR practitioners who want to hire them. As such, it is the most explicitly commercial of the three communities. PR practioners pay $250 in order to post expenses-paid media tours; the writers, who pay nothing to join, apply for the trips. Typical offers vary from free nights at a resort in Maine to a cruise to the Galapagos to a tour of Sri Lanka. "Travelwriters is based on a simple principle," it says, "to connect top-tier writers with editors, PR agencies, tourism professionals, CVBs and tour operators, nurturing the important link that so heavily influences the travel media."[73]

So, to the second question: Is building a travel community site an opportunity for an entrepreneurial travel journalist? Depends on the site, of course, but a single factor appears to be definitive: size of audience. TripAdvisor, for example, is the second most visited travel-related site in the world, according to alexa.com.[74] It generates nearly $400 million in advertising revenue for its parent, publicly held Expedia.[75] The other community sites pale in comparison. Wikitravel, by definition, is free. Regional newsletter travelwriternews. com appears to be a labor of love.

# Mobile: The Largest Opportunity Going Forward

*App* was the American Dialect Society's "word of the year" for 2011.[76] No surprise, given the surge of mobile devices and user appetite for things these devices can do—the apps.

For the travel journalist, mobile applications represent the largest entrepreneurial opportunity going forward. The rapid adoption of mobile technology—by travelers and others—drives the opportunity. The demand for "anytime, anywhere" news, information, social networks, and entertainment has become insatiable.[77] Mobile telephone adoption—about 40 percent of it smartphones—is nearing 100 percent.[78] By 2013 mobile devices will overtake PCs as the most common web-access device worldwide.[79]

Journalists and the organizations they work for have taken note. Newspaper and magazine publishers are adjusting their business models to accommodate the change. A 2010 study by the Audit Bureau of Circulation found that:[80]

- Eighty-nine percent of publishers believe that more people will rely on mobile devices as information sources in the next two years, up from 85 percent a year earlier.
- Eighty-eight percent of newspaper publishers and 57 percent of magazine publishers said they are distributing content on mobile devices, up from 56 and 42 percent a year earlier.
- Sixty percent of publishers said the most successful business model for delivery of content on mobile devices will rely both on advertising and subscription revenue.
- And, in an important nod to entrepreneurs, 49 percent of publishers said they rely on third-party platforms, and another 7 percent said they plan to in the next year.

So, how to get going in this space? There is a lot of accessible, intelligible, journalism-oriented advice, including advice on where to go to learn design and testing, how to sell advertising, and dos and don'ts based on experience.

Start, perhaps, with the advice of Rachel Hinman, a senior research scientist at Nokia's research center in Palo Alto, California. Mobile is a hot, volatile space that is always changing, she says. Recognize what is unique about mobile devices: They are small, always with you, and good for timely information. Understand the context in which mobile devices are used: partial attention and frequent interruption. Focus narrowly, ruthlessly, on helping users get things done.[81]

Learn, as a user, before thinking of becoming an entrepreneur. As director of interactive for the *St. Louis Post-Dispatch*, Will Sullivan understands this thinking. He developed the "Mobile Journalism Reporting Tools Guide"[82] to help journalists learn the hardware and applications. "The best way for journalists to learn is to actually use mobile devices in their reporting," Sullivan told Poynter's Damon Kiesow.[83] Others agree. "Once reporters start to wonder what they can do with mobile," says Clyde Bentley, "they start to get creative."[84] Their advice is certainly applicable to travel journalists. As Sullivan and Bentley recommend, acquire a smartphone and start using the travel-related apps in your work.

The temptation, in the next step, is to begin developing mobile apps. Bentley, among others, discourages this. "The various app markets have become incredibly cluttered, almost unusable," Bentley told the Knight Digital Media Center. "This runs counter to the basic app concept, which is to make it easier."[85] Amy Gahran advises first covering the most important mobile bases: mobile access to your website, email, and texting. Only then consider developing a mobile app.[86]

Some mobile apps are easier to start than others. Virtually no one starts from scratch. Rather, they identify an existing platform on which to adapt or build an application. Many in the travel-related space are choosing geolocation platforms such as Facebook Places, SCVNGR, Foursquare, Gowalla, and Loopt. "Geolocation is the tech buzzword of the year," *PCWorld* wrote in 2010, "and could revolutionize the way we socialize and discover new places."[87]

Most geolocation apps do two things. First, they report your location to others, such as your friends on Facebook or your followers on Twitter. When you "check in" at a bar, restaurant, shop, or museum, your friends know you are there. They, in turn, can recommend what to order, what to buy, and so forth. Second, the apps report your location to the venue, which may reward your presence with points, prizes, and discounts.

Foursquare dominates with seven million users—and counting.

A travel-related geolocation-based mobile app launched by the *Cincinnati Enquirer* in 2010 offers a good case study. The app is called Porkappolis, a play on Cincinnati's reputation as the "slaughterhouse capital of the world."

About the app: Apple's app store describes the app this way: "Porkappolis lets you 'check in' to locations, read reviews, find deals, locate friends, get 'inside information,' earn badges, win prizes, share photos and send updates to Twitter, Facebook and Foursquare. Explore Cincinnati in a whole new way."[88]

How it was developed: The *Enquirer* turned to Double Dutch, a San Francisco start-up focusing on software for location-aware devices. By the end of 2009 Double Dutch had an early version of a platform it was prepared to offer to others. Porkappolis was among its first takers.

How it makes money: Porkappolis is free to users. The money comes from local vendors who typically associate with geolocation apps: bars, restaurants, shops, attractions. "Basic" service is free: The vendor's offer is sent to users who "check in" at the vendor's site. "Enhanced" service is $49 per month. "Premium" service is $149 per month. For that, the vendor's offer is sent to users who "check in" near the vendor's site, the vendor gets 10,000 banner impressions on other *Enquirer*-owned sites, and the vendor gets weekly reports of the number of "check in" users.

The outlook: Too early to tell, but one well-regarded critic is hopeful. "I like it because it's a bold move by Gannett," writes Greg Sterling. "And because it has personality. It's also a new brand."[89]

What's to be learned: There is value to be mined from such hyper-local, geolocation-based apps. An entrepreneur can get into this space more quickly—and more affordably—by contracting with a developer to adapt its app to your market.

A good place to look for other examples of travel-related mobile apps is at appolicious.com. Start with articles tagged "travel." You'll find, for example, reviews of travel-related mobile apps written by freelance writers such as Kathryn Swartz, a University of Missouri journalism graduate, who says she "doesn't know how people lived pre iPhone."[90]

Some recent Swartz articles that suggest the range of travel-related apps on offer:

- "Eat on the cheap with BiteHunter iPhone app's deal finder"
- "Songkick Concerts creates a custom calendar based on your music prefs"
- "Travelers get only the best local spots with 'Best of . . .' iPhone app"
- "Chow down chain free with LocalEats iPad app."

## Using Social Media to Drive Audience Engagement

The entrepreneurial travel journalist succeeds to the extent that she is able to engage an audience. Only an engaged audience will contribute to the journalistic and business outcomes that sustain the entrepreneur—like buying something. So it makes sense to understand what audience engagement means and how to use one set of tools—social media—to drive audience engagement.

Traffic and engagement are related, but not the same thing.

Traffic is the "what" of audience behavior. In print, it's typically a crude quantitative measure of paid circulation, market penetration, or readership. Online, the measures are just as crude: visits, visitors, time on site, page views, bounce rates, sources, and so forth.

Engagement is different. If traffic measures the "what," engagement measures the "what it was like." Avinash Kaushik, a leading thinker about online audience measurement, says engagement is about audience experience.[91] Was it positive or negative? Was it favorable or unfavorable? Did it raise or lower interest? "We should all try to create website experiences that draw favorable attention or interest," Kaushik writes. "The challenge in the context of measurement is that 'favorable attention or interest' is incredibly hard—if not impossible—to measure."[92] One can measure the "degree" of engagement, that is, the extent to which the audience is positively or negatively involved. What cannot be measured is the "kind" of engagement, that is, the emotional state of the audience.

Journalists, like others working online, struggle with what they mean by engagement and how it is measured. Joy Mayer, while a fellow at the Reynolds Journalism Institute at the University of Missouri School of Journalism, divided audience engagement into three categories.[93] They are:

- Community outreach—With this, Mayer embraces the rhetoric of the civic journalism movement.[94] Journalists are engaged with their communities, not aloof. Outreach means identifying needs, sharing expertise, building connections, supporting enrichment.
- Conversation—Here the focus is on journalists "listening as well as talking," then acting on what they hear. Again, in the tradition of civic journalism, Mayer wants journalists to host discussions and participate in discussions hosted by others—both in person and online.
- Collaboration—Ceding power and influence are central to Mayer's concept of collaboration. She uses words such as *soliciting* and *valuing* and *recognizing* the worth of community contributions. "We can accomplish things with the cooperation of the community that we could not do alone," she writes.[95]

Audience engagement in each of these categories can be strengthened—and maybe even measured—through social media. Social media is not one thing, it is many things that share one characteristic: the ability for "anybody to communicate with everybody."[96] Jim Sterne includes seven broad categories in his "social media catalog."[97] They are:

- Forums and message boards—Successors to bulletin boards and newsgroups, forums and message boards are the earliest forms of social content online. Threaded discussions tend to be the most engaging. A travel-related, best-practice example is Lonely Planet's Thorn Tree Forum.[98]
- Review and opinion sites—Most are built on a wiki-like consensus model with outlier opinions washed away. Some, such as Amazon, use this approach to calculate recommendations. Others, such as CNET, offer expertise. A travel-related, best-practice example of the former type is TripAdvisor,[99] and of the latter type, the *Perrin Report* in *Conde Nast Traveller*.[100]
- Social networks—Networks such as the Facebook platform exploit the influence that "friends" have over one another. Facebook and social networks like it permit "friends" to decide how public or how private its communications will be. A travel-related, best-practice example of using the Facebook platform to engage users is Vacation Gals.[101]
- Blogging—Blogs are commentaries posted in reverse chronological order, typically by an individual. Most blogs, including travel-related blogs, combine material gathered by the blogger with material created by the blogger. Comments are the best indicator of user engagement. A travel-related, best-practice example is *Gadling*.[102]
- Microblogging—Microblogs are posts of 140 or fewer characters. Sometimes microblog posts are short messages. Increasingly, they are links without commentary to content elsewhere online. Nearly 80 percent of travel-related microblogs are operated by travel and tourism PR practitioners.[103] Twitter dominates.[104] Travel-related, best-practice examples include Charles Trippy's @CharlesTrippy, Evelyn Hannon's @Journeywomen, and J. D. Andrews's @earthexplorer.[105]
- Bookmarking—Social bookmarking services such as Delicious enable users to organize and share sites across the web. Users "tag" sites with terms such as *travel*. Delicious keeps track and reports the "most popular" on its homepage. When shared, others can search and see bookmarked sites based on tags or people. In mid-2011 there were 2.6 million bookmarks tagged as "travel" on Delicious.[106]

- Media sharing—Tools such as Flickr and YouTube transformed the concept of sharing photos and video. Many others divvy up the media-sharing space today. Quality photos and video are more important than text, leading travel journalists argue.[107] A travel-related, best-practice example is Matador.[108]

As Kaushik and others have pointed out, measuring "engagement" is difficult. Analytics, that is, quantitative measures, only get at the degree of engagement, not the kind. Qualitative measures—such as surveys and focus groups—are needed to understand the kind of engagement.[109]

Journalists increasingly are measuring engagement by how users act upon the information they encounter at a site. Google's PostRank is an organization that measures engagement in these terms by what it calls "engagement events." An event is "where and when stories generate comments, bookmarks, tweets, and other forms of interaction from a host of social hubs."[110]

Outcomes, of course, are what count: more revenue, less cost, higher rates of customer satisfaction. Social media, done right, is a powerful driver of audience engagement.

# An Approach, If Not a Step-by-step Guide, to Get Started

Any step-by-step guide based on the state of the art in technology will quickly be outdated. So this section offers more of a conceptual approach that is likely to endure changes in technology. The approach embraces four concepts. They are:

- Innovate quickly, frugally
- Use free open-source technology
- Match content to audience
- Monitor progress

# Innovate Quickly, Frugally

We noted earlier in this chapter that most innovations fail before they reach the market. Since cost and risk increase with time, the imperative is to "fail fast," before costs mount in pursuit of a bad idea. A new approach to innovation has emerged behind this concept. In some places, it's called "frugal innovation."[111] In others, it's called "lean start-up."[112] Whatever the name, the approach is the same.

It begins with the concept of a start-up as a "temporary organization designed to discover a profitable, scalable business model," says entrepreneur Steven Blank.[113] The idea is to find "the minimum viable product."[114] Most lean start-ups focus on the web, exploiting free, open-source technology that holds down costs.

But holding down costs does not mean innovations that are cheap or second-rate. *The Economist* cites a battery-powered, hand-held electrocardiogram developed by General Electric's innovation lab in India. It's as effective as its plug-in cousin, but one-third the cost. It had to be to succeed in India, a relatively poor country with high rates of heart disease.[115]

A key to the lean start-up is to think small. Facebook began as a messaging service, Blank notes.[116] "Every company wants to hit it big with market-shattering innovations," the *Wall Street Journal* wrote, "but the little changes, too, can make a huge difference."[117] A simple blog about a military museum in one Army town might grow to a full-function website about all U.S. military museums—if the blog proved there is a sustainable audience and revenue stream. The simple blog provides the "early data" so important to convince venture capitalists to invest in the expansion.

There are lots of travel-related examples:

- An early example is the work of Anelia K. Dimitrova, a native of Bulgaria, who launched a no-budget television program in Cedar Falls, Iowa, focusing on international topics.[118] That show, *Here and There*, is now gone, but Dimitrova is still telling internationally focused stories, now on an Iowa-wide blog called *iRovingReporter*.[119]
- When high-end city magazine publisher Emmis Communications sought to test an online presence, it adopted a simple, cheap approach: email newsletters. One of them, in its Atlanta market, is a monthly update on events and travel deals in the Southeast.
- The travel blog *Gadling* partnered with Capital One to promote a travel-oriented credit card rewards program with a sweepstakes, the Most Frustrated Traveler Contest. It takes advantage of a *Gadling* strength—a large audience of travel-savvy readers—and a Capital One strength—a capacity to reward sweepstakes winners with travel.[120]

Two other keys worth noting. One is about the composition of a start-up team. Mark Bowden, whose 29-part 1998 serial "Black Hawk Down" remains a model of journalistic innovation, recommends that a writer partner with a programmer and a visual journalist.[121] Bowden's partner for "Black Hawk Down," Jennifer Musser-Metz, embraced both skill sets. She and her team turned Bowden's "piles of audiotapes, notes, documents, radio transcripts [and] photos" into a web presentation that "is still in the vanguard of Internet presentation."[122] It's rare, even today, that one journalist will have all three skills—thus the idea of a team.

The other key is about saving time. Entrepreneurial journalist Julia Scott, who reports about saving money at bargain.babe.com, says that time is her most important resource, and she can't afford to waste it. She offered three tips about saving time to the Knight Digital Media Center, where she was a fellow in 2009:[123]

- Filter email—Scott uses the Gmail Filters tool to sort incoming messages into three folders. One is for emails from top sources. A second is the advertising requests. A third is for "a whole slew of distracting emails that go straight to archives."
- Reduce the length of your workday—Sounds counterintuitive, but Scott makes herself this increased-productivity deal: "I get to leave one hour early if I finish everything. And I mean everything."
- Adopt an automated calendaring tool—Handwritten to-do lists aren't good enough, according to Scott, because "about half my tasks were the same everyday and many others had to be repeated once or twice a week." The Google Calendar tool (and many others like it) automates the process.

The promise of frugal innovation, of lean start-ups is clear. "If it works, it will reduce failure rates for entrepreneurial ventures and boost innovation," says Thomas R. Eisenmann, a professor at the Harvard Business School. "That's a big deal for the economy.[124]

## Use Free, Open-source Technology

"I never pay for software," says Adrian Holovaty, founder of EveryBlock.[125] Holovaty is not a software pirate. Rather, he is committed to using (indeed, creating) what is called open-source software. The movement, with its origins at the formation of the Internet in the 1960s, refers to a cooperative, often collaborative process of software development. The process requires making the software's source code available to the public, hence the name. Software popularly associated with open source includes Linux, Apache HTTP Server, and Mozilla Firefox.

Whether technically open-source or not, all of the software an entrepreneurial travel journalist needs is free (that is, no cost). Key how-to texts such as *Journalism Next* identify what's free, and how to use it, in three key categories. They are:

- Publishing platforms: WordPress and Blogger have emerged as de facto standards for early-stage publishing. Both have free versions as well as inexpensive "premium" versions.
- Media sharing: Flickr and YouTube are both free. Flickr is for photos, YouTube for videos.
- Analytics: As Avinash Kaushik has observed, a search on Google for "free Web analytic tools" produces 49 million results.[126] His point: no need to pay when so much is free. Google Analytics and Google Insights are good starting points.

One caution: These tools, while free, offer little or no customer service. But you are not on your own. All have online tutorials and various levels of online help. The tools are widely discussed within user groups whose members are generous with advice. All are easy to find and join.

## Match Content to Audience

The starting point is having an idea—ideally, one that matches content with audience. We know from Chapter 3 that the audience for travel journalism is large, varied, and demanding. We know from Chapter 4 that we can reach the audience for travel journalism through four story types and more than a dozen topic niches.

This approach says form and evaluate your idea based on matching the content you want to produce with the audience you want to serve. Most journalists, including travel journalists, begin with the former—content they want to produce. Employed journalists historically haven't had to care much about whether the content they want to produce will find an audience. Entrepreneurial journalists must do so, because the audience determines whether you succeed or fail.

So follow the well-understood, early steps in the innovation process: idea generation, screening and evaluation, development, and testing. An innovation process developed for journalists by the American Press Institute is a good place to start.[127] It begins with a question: What jobs do people want done that we can do well? Answering this question is the point of this approach. What does some or all of the audience for travel journalism want done? Can we do it well?

## Monitor Progress

Once up and running—with analytics tools installed—you'll discover a paradox that hinders monitoring progress. It's the paradox of too much data, too little insight. The analytics tools supply the data—but rarely much insight. But the good news is that frustration with this paradox is so widespread that smart people give it a lot of attention. Here are some ideas:

- Be clear about your objectives, what you are trying to accomplish. Is it enough that an audience reads your content? If not, what do you want the audience to do? Comment, recommend, buy? Once settled, these are your "key performance indicators."
- Ignore data that doesn't bear on your objectives, on what you are trying to accomplish. Instead, focus on data that shows the direction of your key performance indicators. This is your "dashboard."
- Consult your dashboard periodically. Don't be whipsawed by the moment-to-moment ups and downs. What does the dashboard show monthly, quarterly, yearly?

Ultimately, return on your investment—broadly understood—is the only thing that matters. What did you get for the time and money invested? If your objective is to establish a stand-alone enterprise, is the return adequate to sustain your effort? If your objective is to prove up an idea, is the return sufficient to convince investors to buy, expand, replicate? The dashboard of key performance indicators will help you monitor progress along the way.

## So Where to Start? Try a Simple Blog

The travel blog is a well-understood, tested medium. There's no better place for the entrepreneurial travel journalist to start. Observe the advice of the freelance travel journalists I surveyed and start with a local topic, remembering that "almost everyone's 'backyard' is a destination to someone else," as one freelancer said, and understanding that it's a longer-term opportunity to develop expertise, a reputation, a niche.

Experts[128] suggest these steps:

- Set a goal for the percentage of content you intend to aggregate and the percentage you intend to create. Some bloggers aggregate as much as 80 percent of their content, but as the "expert," your percentage will probably be much lower.
- Select a platform, either Blogger or WordPress.
- Set up RSS feeds to provide information and links. The challenge is identifying the keywords that will harvest salient information. Study the keyword analysis on alexa.com for advice.
- Connect to the audience through comments, Facebook, and Twitter.
- Test various revenue streams: Google's AdSense is simple to implement and aligns well with Blogger and WordPress. Adopt one of the micropayment apps, such as PayPal's Digital Goods.
- Create and monitor a dashboard of key performance indicators. Blogger and WordPress provide simple visitor and search data. Use a free analytics tool such as Google Analytics for more granular, useful data.
- Add simple multimedia, such as photographs using Flickr and video using YouTube.

## Be Aware of Legal, Ethical Issues

Entrepreneurial travel journalists confront both legal and ethical issues. Unlike their better-heeled, employed-by-large-organization colleagues, entrepreneurial journalists lack legal resources. Indeed, start-ups have failed because entrepreneurs could not cover their legal costs.[129]

The Citizen Media Law Project at Harvard's Berkman Center for Internet and Society is an excellent, free resource.[130] So is the Reporter's Committee for Freedom of the Press, though it is more focused on access to information than the legal issues associated with online publishing.[131]

At a minimum, get smart about the following two issues. The first is liability for publishing defamatory statements made by others. The Communications Decency Act governs this. The other is fair use and infringement of copyright under the Digital Millennium Copyright Act (DMCA).

There are ethical as well as legal issues around aggregation or, as some call it, *curation*. "Critics of the curation economy tend to fall into two categories," writes Steven Rosenbaum, "those who say it's immoral and those who say it's illegal."[132] Fair use and copyright govern what's legal. Ethics govern what's moral.

The DMCA provides the legal framework for aggregation. Aggregators typically publish DMCA-compliance notices on their sites, some more prominently than others. See, for example, the policies and notices posted by entrepreneurial journalist Julia Scott, who blogs as Bargain Babe.[133]

The ethical framework is a clamorous debate. "The debate seems to be drawn across content-creator party lines," Rosenbaum says. "Old-school makers see sharing as theft; new-school bloggers see the linked economy as a bonanza waiting to be monetized by forward-thinking media companies. It's hardly a balanced debate."[134]

## Test Your Understanding

1. Why is "mindset" key to the success of the entrepreneurial travel journalist?
2. Why are journalists "disadvantaged" when it comes to entrepreneurship? How might journalists overcome these disadvantages?
3. Why is mobile thought to be the largest opportunity for the entrepreneurial travel journalist?
4. What is meant by "audience engagement?" How can social media be used to drive audience engagement?
5. Describe the four-step approach to getting started as an entrepreneurial travel journalist.
6. Imagine and describe the content of a simple travel blog for your market.

## Practice Your Skills

Assignment #6—Develop a simple website based on the story outlined in Chapter 4. Follow the steps recommended by experts discussed in this chapter, as follows:

1. Set a goal for the percentage of content you intend to aggregate and the percentage you intend to create. Some bloggers aggregate as much as 80 percent of their content, but as the "expert," your percentage will probably be much lower.

2. Select a platform, either Blogger or WordPress.
3. Set up RSS feeds to provide information and links. The challenge is identifying the keywords that will harvest salient information. Study the keyword analysis on alexa.com for advice.
4. Connect to the audience through comments, Facebook, and Twitter. Test various revenue streams: Google's AdSense is simple to implement and aligns well with Blogger and WordPress. Adopt one of the micro payment tools, such as PayPal's Digital Goods.
5. Create and monitor a dashboard of key performance indicators. Blogger and WordPress provide simple visitor and search data. Use a free analytics tool such as Google Analytics for more granular, useful data.
6. Add simple multimedia, such as photographs using Flickr and video using YouTube.

## A Closer Look: The Diary of an Entrepreneurial Travel Journalist in Training

I became acquainted with Mary Bergin through email. She was generous enough to reply to my survey of freelance travel writers who are members of the Society of American Travel Writers.

I wanted to follow up with her because, more than most of her colleagues, Bergin exhibited a keen understanding of entrepreneurial journalism. Her website, roadstraveled.com, exploited three key revenue opportunities: advertising, product sales, and fundraising through sponsorships.

"Can you tell me a little more about your revenue strategy," I asked, "how it developed, which channels do the best, and so forth?"

Bergin's reply was a surprise:

"Truth be told," Bergin wrote, "very little of my revenue comes from what you see online."

Indeed, her living still depends on her travel writing and photography sold to newspapers and magazines in print. "I also get book royalties, do guidebook fact checking and public speaking, plus a little bit of consulting work," she wrote.

Bergin retains the copyright to all material and repurposes it on roadstraveled.com "because I know an online presence is becoming increasingly important." To that end, Bergin applied to the Poynter Institute for Media Studies for its week-long seminar on digital entrepreneurship, sponsored by the Ford Foundation.

Bergin was accepted and, as it turns out, so was I. I asked Bergin to keep a diary of her impressions for this book.

Herewith, her entries from the week at Poynter. Each begins with her post on Facebook, followed by reflections from the seminar.


**Pre-seminar Expectations:**

"I seek to refine my business strategy and online presence to allow for smart, efficient, ethical and personally fulfilling growth that will continue to earn me a living wage in writing and photography. Since 2002, my target audience has been people who read my weekly travel columns in daily Wisconsin newspapers. Combined print circulation peaked at 300,000 from a dozen dailies. Hard to know how much online exposure (through customers' Web sites) replaces print circulation decreases."

Thanks to the Internet, now my audience potential expands beyond newspaper subscribers/buyers. The target today is people who live in Wisconsin, enjoy travel in Wisconsin or maintain another type of allegiance to Wisconsin (because of family, friends, schooling, etc.).

I come to Poynter wondering whether my digital niche is too narrow, my approach too old-school, my path realistic for providing a living wage—as traditional writing and photography outlets shift and dwindle.


**Monday's Facebook Post:**

"Invigorating mix of Poynter classmates: staffers from *The Nation*, ABC in NYC, WGN in Chicago, social issue activists, community journalists. All 20 of us have digital ideas to run with, but we're not sure how."

Thrilled to see that I'm not the only student in midlife here, and that my work in travel journalism is not the only lifestyle endeavor within this sea of hard news devotees.

We are introduced to the notion of building a strong team in our work. Up until now, I've pretty much operated as a one-woman band who eyes like-minded media as competitors, not partners.

**Tuesday's Facebook Post:**

"Many lessons today. My favorite: Dig for the pain. If there's no problem, there's no need for a solution."

Twisting my perspective from "it's all about me" and my career goals/survival to "what do you need and how can I help?" is vital in plotting my future path. If I can pinpoint why and how my work is of value (or, better, indispensable) to others, I'll better understand the most logical way to make progress.

It's another adjustment that sort of seems like ministry. I've spent a lifetime thinking like the average newspaper reader while trying to please an editor and his/her priorities about newsroom coverage.

Now, to thrive in this digital world, I need to be "relentlessly focused and useful," especially to the most influential voices online—which can be measured like anything else, thanks to wefollow.com.

**Wednesday's Facebook Post:**

"Advice worth two asterisks today: This is no time to run a private coffee shop. Reach your audience where they are. Don't force them to come to you."

Pleased to find out that a smart and insightful instructor here, CUNY's Jeremy Caplan, also teaches an online intro to video class for SATW this month. A nice coincidence, since I signed up weeks ago. Just hoping to find the time, energy and concentration to learn one more major skill in this little window of time.

Life proceeds more and more like this. Digital skill sets need to expand, but deadlines for projects that pay the bills take priority, right?

"Start small," we are advised, then divide, conquer, make mistakes and learn from them.

I am surrounded by risk takers and find this invigorating.

**Thursday's Facebook Post:**

"True confession: I am a digital hoarder, saving almost everything but not utilizing what I have in a smart way. Time to expand/update my email contacts list and work it."

OK, that's what I would have posted, had I not gotten myself so immersed in homework, one-on-one chatter and another packed day of lectures.

Where's the money to leverage these great ideas? In a perfect world, advertisers arrive when online traffic increases. Traffic increases when content is of value and changes often. How to expand on quality coverage? I'd go nuts trying to do everything alone, but maybe I can add links to voices and relevant content other than my own.

Many voices produce great ideas at Poynter. The notion of helping small businesses design simple ads or find a voice in social media seems intriguing. Same for creating unique travel experiences that have commercial potential. No need to think too small or box myself in too tight.

Coming next, and tonight, over dinner: discussions about survival and ethics in this particular field of journalism, which is full of land mines that test integrity and resilience. To me, the conversation begins with questions: Is travel writing of value? If "yes," why—and who should pay to make the path more ethical?

I've been asking myself these questions since beginning this work in 2002.


**Friday's Facebook Post:**

Hah! There is none. Too consumed with perfecting my 60-second business pitch to my peers and a panel of faculty. It goes like this:

"Hello. I am Mary Bergin and I come from the land of cheese, beer and Green Bay Packers. This is who we are in Wisconsin, but what I do is travel beyond the obvious, literally, and since 2002 I've taken at least 200,000 newspaper readers along for the ride, every week, through my syndicated travel columns about the heart of Wisconsin.

"Now RoadsTraveled.com expands the insider advice and connections. It will not tell you what you already know. It will not be a laundry list of all attractions open for business. That's what state tourism and community websites do, because they have to treat all their kids—their affiliates—the same.

"I sniff out the unusual.

"When I shop for used books, I head to a converted manure storage tank.

"I order pizza at the ruins of a barn near the Mississippi, where what you order outdoors is made from ingredients harvested just yards away.

"I find beer and camaraderie at Wisconsin-centric bars in Manhattan and Frankfurt, Germany.

"My online traffic has tripled in two years. Upcoming reader incentives should further boost it, and as numbers grow, so will advertising, especially from rural entrepreneurs who know we share their voice and values.

"More than travel, this venture is all about who we are in Wisconsin, and if you share that pride and identity, you'll want a stake in it."

I ask faculty to point out the gaps and red flags. Sources of financing will be crucial, and I need to reach Wisconsin travelers who are not from my home state.

It's been a full and satisfying week. I scribble to-do lists, priorities and organize tasks on the plane ride home.

## · 7 ·

# WHAT ARE THE FUNDING OPPORTUNITIES FOR THE TRAVEL JOURNALIST?

A button on the main navigation bar at the travel-advice website thevacation-gals.com is both unexpected and welcome.

It reads, "Disclosure Policy."

Click on the button to find this: "As is common in the travel industry, we may receive complimentary travel or reduced rates at hotels, resorts and attractions. We disclose such offers in our posts, and we require our guest bloggers to disclose in their posts."[1]

The "Disclosure Policy" button is unexpected because it is rare. Most travel journalists and their publishers do not disclose to their readers that they have received free or reduced-rate travel. According to those who responded to my survey of freelance travel journalists, for example, 90 percent who participated in 2010 media tours accepted free or reduced-rate travel, but just 20 percent disclosed it in travel blogs and websites they control.[2]

The disclosure policy is welcome because it resolves, for many, a conundrum facing most travel journalists: how to fund an essential element of the work—the travel—without compromising journalistic standards.

Funding the cost of travel is at the center of most debates about travel journalism.

As early as 1977, Sigma Delta Chi sent a tough-minded letter to the Society of American Travel Writers criticizing the practice of SATW members taking free trips. The SATW's president, Ben Carruthers, replied, in effect: Mind your own business—we have no practical choice.[3] Nearly a quarter-century later, another protector of journalistic standards made the same case to the SATW—and got the same response.[4]

Conflict of interest is embedded in the way most travel writing is produced. The process—often referred to as subsidized travel, press trips, or media tours—has evolved over six decades. It works like this:

- Marketers want editorial coverage of their destination because it is more influential than advertising.[5] They offer to pay the expenses of travel writers willing to visit and write about their experiences.
- Travel writers accept the deal because they can't otherwise afford to visit the destination. Why? Because the rates most publishers pay for the work are too low to cover the cost of travel and provide a living.
- Publishers go along because it keeps their costs low.
- Readers are kept in the dark because the free-travel-in-return-for-coverage is rarely disclosed.

Neither the process, nor the criticism of it, is new.

Newspapers and magazines have reported about travel since the 1930s, but most merely draped PR material around their ad-filled travel pages. Subsidized travel emerged in the 1950s, around the time jets opened the world to travel as "a basic tenet of the American Dream."[6] Marketers wanted to expose travelers to this new mode of travel and the destinations it could reach—and were willing to pay for editorial coverage and advertising. As millions more traveled, publishers benefited—travel-related page counts, advertising linage, and revenue rose—but travel desks didn't—neither staffing nor budget for travel rose as fast as revenue.[7] How to fill the burgeoning number of pages? Almost immediately, freelancers emerged to fill the gap, and with them, questions emerged about expenses, rates, and disclosure. Over time, most travel desks accepted free or reduced-rate travel for freelancers, some did not, and others ignored the question.[8] Little about the process has changed. It is, Folker Hanusch observes, one of the "special issues" that distinguish travel writers from journalists.[9] If anything, the process has been codified: The Society of American Travel Writers permits its members to accept free or reduced-rate travel so long as they disclose the fact to their editors—not to readers.[10] And that's the industry's strictest code.

Indeed, the process is so entrenched that some of the dominant "how-to" travel-writing texts contain instruction on "how to get in on this action," as one promises.[11] Some of these texts, at the same time, are clear—even forceful—about the pros and cons of accepting free or reduced travel. "Both press trips and freebies introduce the thorny issues of objectivity and impartiality," writes Don George in *Travel Writing*. "If a hotel has given you a free room, how objective can you be in assessing it and writing about it?"[12]

As early as 1973, the *New York Times* explored the process with a travel-book author who referred to writers who accept free or reduced-rate travel as "freeloaders" and "whores."[13] The language and tone of that criticism has showed up regularly since. The *Columbia Journalism Review* weighed in with a 1974 story titled "The Fantasy World of Travel Sections." Written by *Wall Street Journal* reporter Stanford N. Sesser, the story asserted: "But now with travel such big business and with millions of Americans visiting distant points, it becomes relevant to ask why travel can't be reported in a considerably more professional and probing fashion than it is."[14] A panel of top editors from the American Society of Newspaper Editors concluded in 1988 that "editors must take travel editor and writers 'off the dole' of the travel industry, establishing clearer ethical standards."[15] As recently as 2000, the *American Journalism Review* called subsidized travel "a profoundly bad idea, with conflict of interest written all over it."[16]

Some publishers follow a different process. They pay the expenses of travel journalists to whom they assign work. They don't accept work from travel writers whose expenses were paid by marketers. Three travel magazines and associated websites that follow this practice are *National Geographic Traveler*, *Travel + Leisure*, and *Arthur Frommer's Budget Travel*. "We do not accept proposals about trips that are subsidized in any way," *National Geographic Traveler* tells writers in its Writers Guidelines.[17] *Travel + Leisure* promotes its policy in the magazine. "*Travel + Leisure* editors, writers and photographers are the industry's most reliable sources," it says in its Editor's Note. "While on assignment, they travel incognito whenever possible and do not take press trips or accept free travel of any kind."[18] *Budget Travel* also explains its policy in the magazine: Next to a bright yellow starburst that reads "Why you can trust us," its editor writes, "*Budget Travel* writers don't accept free trips or discounts—because it's impossible to be objective if someone else is paying the way."[19] It is worth noting, perhaps, that *National Geographic Traveler*, *Budget Travel*, and *Travel + Leisure* rank first, second, and fourth, respectively, in the number of times they've won the Lowell Thomas Award for top travel magazine since 2000.[20]

Travel journalists—especially freelancers—are not of one mind about this. I surveyed three groups of travel journalists on the subject of accepting and disclosing to readers free or reduced-rate travel. They are business journalists who cover travel, editor members of the Society of American Travel Writers, and freelancer members of SATW.[21]

Among the three groups, business journalists who cover travel are the most opposed to free and reduced-rate travel and the most in favor of disclosure. Some 89 percent who responded to my survey somewhat agree or strongly agree that journalists covering travel "should never accept anything of value from a travel or tourism provider." Some 78 percent strongly agree that these journalists "should disclose to readers and viewers anything of value they accept from a travel or tourism provider."

Just 18 percent of SATW editors who responded to my survey somewhat agree about not accepting anything of value; none strongly agree. But 59 percent of the SATW editors somewhat agree or strongly agree that accepting anything of value should be disclosed.

Only 11 percent of SATW freelancers who responded to my survey somewhat agree or strongly agree about not accepting anything of value. Indeed, 51 percent strongly disagree. And 49 percent of SATW freelancers somewhat agree or strongly agree that accepting anything of value should be disclosed.

The data make sense. Most of the business journalists who cover travel and tourism are employed by newspapers, reporting to business editors. They cover travel and tourism as a local business news story, much as they would cover major employers, retailers, or service industries. Their work involves little or no travel. The SATW freelancers, on the other hand, are not employed—they are entrepreneurs. Their livelihoods depend on pitching ideas to travel editors. They are two to three times more likely to cover destinations than news, and their work involves much travel.[22] The SATW editors are somewhere in between.

Many travel journalists, and especially the freelancers, reject the assertion of conflict of interest. "My opinion is not for sale," is the common refrain. The *American Journalism Review*'s Rem Rieder raised the issue before a 2000 gathering of SATW members in Wales. "Many of the travel writers took the view that they are professionals who call them like they see them, who would never be swayed by a free hotel room or a luxury cruise," Rieder recalled. "They saw no merit in that age-old bete noir, the appearance of a conflict of interest. They seemed to find my concerns on that score a little quaint."[23]

Some freelancers frame personal guidelines with respect to free and reduced-rate travel. "To keep the obligations/ethics situation in bounds,"

Carol Barrington decades ago adopted this three-way test: "The sponsor of the trip is buying my time and attention, but not my opinion. I guarantee only that a story will be produced from a trip and that I will attempt to market it. There is no review of manuscript."[24]

Freelancers note that journalists on other beats accept things of value without the same criticism. John Hulteng traced early examples in an ethics guide for ASNE: sports writers who serve as paid scorers at baseball games, editorial page editors who accept government-paid trips to countries such as Israel, and photographers who are lent exotic gear by camera and lens manufacturers.[25] Others have included book critics who receive review copies and theater critics who receive free tickets. More recently, sports journalists have signed endorsement deals with shoe manufacturers.[26] And, then there are the auto writers— they are famous for gobbling up freebies the manufacturers offer them at their auto shows. "This has been a great year for swag," one auto writer told the *Detroit News* after walking through the North American International Auto Show in 2011.[27] Auto writers also accept loans of cars they want to review. They return them, of course, but sometimes the worse for wear. Auto writer Peter Cheney borrowed a $180,000 Porsche and parked it in his garage, only to discover later that his teenage son had fired it up and backed it into the garage door. Damage: $11,000 paid by Porsche, who refused Cheney's offer to pay.[28]

It's not true, however, that taking freebies on nontravel beats has escaped criticism. As a result, ASNE discontinued the elaborate dinners sponsored by auto manufacturers at its annual meeting.[29] A book critic for the *Philadelphia Inquirer* was fired after he sold review copies.[30] And the sports journalists with the shoe endorsements were said by one critic to be "on the take."[31]

Some freelancers have accepted free and reduced-rate travel and been negative about their experiences. "Not all travel journalism is tourism's handmaiden," says Lyn McGaurr, a travel journalist and academician.[32] She cites the work of three travel journalists—Jeff Greenwald, Paul Miles, and A. A. Gill—who at different times toured Tasmania under the auspices of a government tourism program, and all three criticized government forestry practices. In a piece for the travel section of the *Sunday Times*, Gill wrote that the island's forests were "being rubbed out by special pleading, arm-twisting and back-scratching corruption."[33] Similarly, a Turkish tour company invited travel journalist Carol Lazar on an expense-paid trip to Istanbul. Among other criticisms, her story mocked "the Shylockian greed of its merchants." She never got invited anywhere else by that tour company.[34] Travel journalist Tim Leffel is annoyed by "the self-appointed ethical experts" who contend that free or reduced-rate travel results in positive coverage. "In my two decades of experience," he says, "this just isn't supported by the reality of what gets printed."[35] Leffel says he once challenged a critic to read 20 hotel reviews he wrote and determine which hotels paid for his stay and which did not. "She flunked miserably, getting one of the hosted ones right out of five and guessing that four had put me up that really had not."[36]

Of course, a handful of freelance travel journalists make a point of not accepting free or reduced-rate travel. One is the *Independent*'s Simon Calder, known to readers as "The man who pays his way."[37] Calder's decision to not accept free or reduced-rate travel is more a matter of journalistic method than ethical behavior. Paying his own way means traveling on the cheap. The cheaper he travels, the closer he gets to the people and the place. "As a result of this somewhat curious and eccentric policy," Calder writes, "I tend to meet a lot of very interesting folk. The people with the best stories to tell live life in the cheap seats."[38]

Another handful of freelance travel journalists have confessed that free and reduced-rate travel influenced their coverage. In a nuanced essay "exploring the ethical swamp of travel writing," Elizabeth Austin confessed: "I was a travel whore."[39] In a confessional piece for the *Columbia Journalism Review*, Jeremy Weir Alderson concluded that "most travel writing simply dishonors our free press."[40] More recently, travel journalist Chuck Thompson issued a book-length confession: "Smile When You're Lying: Confessions of a Rogue Travel Writer."[41]

All this begs the question framed by McGaurr: "How should journalists navigate this issue?"[42] No one believes that travel marketers will stop offering free or reduced-rate travel, or that all travel journalists will stop accepting it. It's too embedded. Many, however, agree that disclosure is the key. "Just tell the reader who picked up the tab," travel journalists say.[43] Simple, common sense. But as Stephen Covey famously observed, common sense isn't always common practice.

Disclosure is not common practice. Neither SATW editors nor freelancers uniformly agree that travel journalists should disclose free or reduced-rate travel—despite SATW's disclosure-oriented code of ethics. Though 59 percent of SATW editors who responded to my survey strongly agree or somewhat agree, 24 percent neither agree nor disagree and 18 percent strongly disagree or somewhat disagree.[44] Among the SATW freelancers who responded to my survey, 31 percent strongly disagree or somewhat disagree. Disclosure, some note, is not within the travel journalist's control. And there is evidence that publishers and editors have deleted disclosures from travel journalist's drafts.[45]

But it's also true that travel journalists fail to disclose free or reduced-rate travel when it's entirely within their control. Some 66 percent of the SATW freelancers who responded to my survey participated in one or more expense-paid media tours in 2010. Some 42 percent of them publish their work on blogs or websites they own or control. Yet, when I read their blogs and websites I found no evidence of disclosure, in any form, in 80 percent of them.

Why make disclosure common practice? Credibility with readers is the most important reason. Readers do not like to be deceived. For an earlier study, I consulted the research on advertorial products—advertisements in the form of editorials.[46] Turns out readers don't object to reading a Lawn and Garden special section with Honda products on the cover so long as they know Honda paid for the product placement. By inference, readers might not object to a destination piece about a luxury resort in Bali so long as they know the travel journalist's stay was comped. "A reader shouldn't have to wonder whether a story is being distorted by financial pressure or whether the newspaper is beholden to special interests," says former *Seattle Times* travel editor John B. MacDonald.[47] Better to not accept free or reduced-rate travel, MacDonald believes.[48] But with disclosure, at least the reader needn't wonder. "If the public's trust in what they see, hear and read is to be maintained," says ethicist Karen Sanders, "the motives and interests behind the news should be as transparent as possible."[49]

How might travel journalism do a better job codifying disclosure of free and reduced-rate travel? Three ideas emerged from the research. They are: First, adopt the "Uniform Disclosure Rule" proposed by Alexander Eliot; second, require travel journalism award entrants to disclose subsidized travel; and third, adopt a "product testing" approach to Service/Advice-type stories.

# Adopt the "Uniform Disclosure Rule" Proposed by Alexander Eliot

Eliot proposed the rule in a 1994 essay in *Editor & Publisher*.[50] "Every published travel piece," Eliot wrote, "should carry a prominent list of acknowledgements letting the reader see who provided what amenities to the author."

The acknowledgements should be "proud and unashamed," Eliot wrote, akin to those scholars write to acknowledge helpful colleagues and institutions. "To be listed thus would garner valuable publicity for the sponsors," Eliot wrote. "Plus, it would provide a reality-check for readers to apply. Has the author provided kid-glove treatment to certain sponsors, or not?"

How might this work? Not as a matter of law, Eliot emphasizes, but "as a norm of editorial practice." This would be no different than any of the other disclosure policies common in the best news organizations. Over time, the rule would be taught in travel journalism classes, embraced by travel journalism associations, memorialized in codes of conduct.

Look to travel journalism in Australia as a model. Australia is farther ahead of other English-speaking countries, including the U.S., with respect to disclosure. It is common to see disclosures at the end of travel stories. Susan Kurosawa, the travel editor at *The Australian*, says the newspaper has a policy that requires full disclosure of travel sponsors.[51] A 2009 study concluded that disclosures appeared at the end of 40 percent of *The Australian's* travel articles, the highest disclosure rate the study observed among three national newspapers.[52]

In print, disclosure is a simple matter of appending a statement at the beginning or the end. It's just as simple online, especially when you adopt a free disclosure widget such as disclosurepolicy.org, from publicly held social media marketer IZEA. Debi Lander is an example of a travel journalist who has adopted the widget, to great effect, she says.[53]

# Require Travel Journalism Award Entries to Disclose Subsidized Travel

Bodybuilding competitions explicitly distinguish between bodybuilders who take growth hormones and those who do not. Might this, in some way, be a model for travel journalism awards?

Harsh as it may seem, subsidies are to travel journalists what hormones are to bodybuilders: For some, they are "a necessary tool of the trade," as one travel editor wrote.[54] For others, they are anathema. Bodybuilders, when they compete, disclose whether they use hormones. Travel journalists, when they compete, should disclose whether they accept subsidies.

Start with the Lowell Thomas Awards, recognized as travel journalism's Pulitzer Prizes. They are administered by a foundation and judged by faculty at a leading journalism school. Entrants are not required to disclose to the judges whether the travel was subsidized, or whether the subsidy was disclosed to readers.

They should be required to do both.

At one level, it's simple transparency, no different than the letters that accompany Pulitzer entries, describing how the entry was produced: What was the influence of deadlines? How were key documents procured? Why were sources granted anonymity? At another level, subsidized travel could become a judging criteria. Might judges give greater weight to entries that did not involve subsidized travel? Might judges give greater weight to entries that did involve subsidized travel, but disclosed the subsidy to readers?

Many travel journalists assert that subsidies do not influence or distort their work. Let's test this: Disclose the subsidies and judge whether they affected the work.

## Adopt a Product-testing Approach to Service- and Advice-type Stories

Travel journalist Matt Kepnes wanted to test whether rail passes produce the promised savings. When planning a trip through Europe, Kepnes asked Eurail for a pass to test. The business school graduate, who blogs as NomadicMatt, said a thorough cost-benefit analysis was his goal.

"I did the trip, and then I wrote down the prices it would have cost without the pass, and I compared the two," Kepnes said in a 2009 interview with Matt Gross.[55] "Did I save money, or did I not save money?"

"Did you?" asked Gross.

"I did," Kepnes said, "but only because I took a lot of long train rides. If you take short train rides, those Eurail passes are not worth it."

This is an example of a product-testing approach to Service- and Advice-type travel journalism. Kepnes is explicit about his motive: He wants to test a product or service that may or may not be a good buy for travelers. He is explicit about his method: a cost-benefit analysis that grows out of business-school training. And he is explicit about his means: Kepnes discloses everything to his readers.

The approach is akin to the automotive journalist who borrows a car to test-drive, or a book critic who asks a publisher for a "review copy," or a food journalist who asks two dozen manufacturers for a sample bottle of their hot sauce.

Product reviews are a fast-growing segment of business online and are something of a mishmash ethically. At the high end are the well-known, long-in-business firms such as Consumer Reports. At the low end are the product-review sites that are fronts for profiting from sales of the products reviewed. You know you've reached the latter when the review concludes with an "order" button.

For travel journalists, here are some possible guidelines:

- It is fine to borrow and return gear, or to receive consumables for free, for the purposes of a test.
- Be clear with the provider about your motives, methods, and means: You're an informed neutral, testing and reporting results—whatever they may be.
- You should have no financial stake in the product review: Do not accept payment from a product provider or a commission from a product sale. It is fine to include "where to buy" information—readers value it.
- Disclose everything to readers.

Taking a product-testing approach to Service- and Advice-type travel journalism is consistent with high journalistic standards. In his study of market-driven journalism, John H. McManus lists seven story topics that represent high journalistic standards. One of them is "wise purchasing of goods and services."[56]

Each of these ideas requires change in attitude and behavior. Anyone familiar with the literature on change management knows that making change is difficult—often impossible.[57]

But the literature also says there are well-understood ways to think about change.

One key, for our purposes, is recognizing that the issue of subsidized travel and the disclosure of subsidies need not be either/or with consequences in each direction. So many ethical issues are framed this way—as dilemmas. Either the travel journalist refuses subsidies, so his travel is limited but he is judged ethical by traditional journalistic standards, or the travel journalist accepts subsidies and therefore travels widely, but he is judged unethical.

The three disclosure-related practices discussed here embrace what ethicist Anthony Weston calls "the need for inventiveness in ethics."[58] The subsidies-or-no-subsidies debate encompasses what psychologists refer to as "set" thinking. The key to resolving the debate, Weston writes, requires "breaking set," that is the habits and assumptions we've long relied on.[59]

One approach he recommends comes from problem-solving expert Edward De Bono. Start by imagining the perfect solution. For our purposes, maybe it's a new business model that funds travel journalism without subsidized travel. Impossible, you conclude. "But don't stop there," Weston writes. "Work backward slowly from the perfect-but-impossible to [what De Bono calls] 'intermediate' solutions."[60]

So, think of the three disclosure-related practices as just that: intermediate solutions that get us past the perhaps impossible-to-solve dilemma of subsidies or no subsidies.

## How to Fund Travel Journalism

With this debate as a backdrop, let's turn to ideas about how to fund travel journalism from traditional and nontraditional sources.

As a starting point, recognize that travel journalists aren't all the same when it comes to funding their work. For some, it's a hobby-level interest. For others, it's a short-term, right-out-of-college experience where it's fine to sleep on couches and eat noodles. For some, it's an adjunct to other, perhaps well-paid work. For still others, it's something for retirement, supported by a pension, Social Security, or a retirement account.

But then there is the journalist who's picked travel as a beat and wants to make a career of it. The question for this journalist is this: Are there traditional or nontraditional ways to fund this work, according to the identity, purpose, and methods I chose to embrace? Can I make a living doing independent, ethical, substantive work?

The answers, alas, come hard.

Traditional sources are both well understood and in decline. They include employment as a travel journalist by a newspaper Travel section, a travel-related magazine, or a guidebook publisher—in print or online. They also include freelancing, the more common, if less lucrative traditional source.

Employment of journalists, including travel journalists, is in steep decline at the 1,400-plus daily newspapers in the U.S. Losses of 5,000 jobs annually have been common since 2007.[61] Today, U.S. newsrooms employ the same number of journalists they employed in the 1980s.[62] The outlook is unpromising. The Bureau of Labor Statistics projects a 6 percent decline in reporting jobs through 2018.[63]

The AJR's Kathryn S. Wenner first noticed a "seismic shift" in travel journalist employment in 2001, when she reported the departure through buyouts of four leading travel editors.[64] "This is really an earthquake in our quiet little corner of journalism," Gary Warner, travel editor for the *Orange County Register*, told her. In Florida, a state prominent for travel journalism, three top travel editors—in Miami, Fort Lauderdale, and St. Petersburg—left their roles since 2001, two due to buyouts.[65]

Some 60 percent of the travel editors who responded to my survey reported employing no writers. They work in newspapers, magazines, and guidebooks—in print and online. Another 18 percent reported employing just one.[66]

This isn't a surprise for the magazines and guidebooks: They rarely employ writers and photographers, depending mostly on freelancers. Freelancing is the most common traditional source of funding for travel journalism. Reliable data about it, however, is scarce.

"Some people make a living as a [freelance] travel writer," writes Tim Leffel. "They are a very small minority."[67] Leffel ought to know. He's written hundreds of travel stories, three travel books, and one of the most recent "how-to" texts, *Travel Writing 2.0*. Leffel interviewed 52 travel journalists for the book. One question he asked was what they earned. Some 40 percent earned $25,000 or less.[68]

Tom Brosnahan, an expert on guidebook publishing, estimates that a writer who gets $30,000 to write a 350-page guidebook will net $6.09 per hour after expenses and taxes.[69] "Every guidebook author," Brosnahan writes, "has stories of guidebook projects that didn't pay."

Better, perhaps, to consider nontraditional sources. Four nontraditional sources stand out, each one requiring some level of entrepreneurship.

They are:

- Self-funding
- Grant funding
- User funding, directly through fees or indirectly through advertising
- Investor funding

Let's look at each source with examples of each in play.

## Self-funding

Self-funding is the simplest: Save up the money, pay your own way. This funds the travel; it doesn't fund the living. The living comes from your "day job," as they say. Self-funding is the approach for travel journalism as a part-time adjunct to something else.

Pam Mandel is a good example. She earns a living as a writer, but not as a travel writer. "I would be unable to pay my bills were I to pursue that line of work full time," Mandel explains on her blog, *Nerd's Eye View*. "I do it on the side because I enjoy it."[70]

Self-funding usually means traveling on the cheap. Simon Calder says that it makes him a better traveler—and travel writer: Learning how to navigate public transport, to shop for food in local markets, and to arrange stays in local guesthouses encourages conversation with locals. (This was true in my experience, too: I traveled to London on the cheap so often— and so effectively—that I developed a travel website, frugal-london.com. Its premise is that everything worth seeing and doing in London is either cheap or free.)

Looking for a funding model? Switch from Starbuck's lattes to Burger King's Seattle's Best, owned by Starbucks, and save about $1,000 a year. Reduce your cable service to "basic" and save about $400 more. That savings alone will pay the roundtrip fare to (almost) any destination on earth, or two or more airfares if you travel to nearby destinations.

A further point is that self-funding permits you to get started in an independent, ethical, substantive way: travel, report, post. In this sense, self-funding is a first option, not just a last resort.

## Grant Funding

Travel journalists, like writers in other genres, benefit from grants, fellowships, and scholarships. Much of Preethi Burkholder's work in Sri Lanka, as an example, has been funded by grants. "Grants offer key sources of financial support for travelers," he writes in a guide to finding travel grants. "Winning a grant is an inexpensive way to raise funds to enrich your personal, professional, and spiritual life."[71] Travel grants aren't just for experienced travel journalists. Many are for college students, recent graduates, and just-starting-out professionals. A 2008 guide by well-traveled radio journalist Emma Jacobs lists 31.[72]

There are two underutilized grant opportunities to fund travel journalism. One is the grants that fund independent foreign correspondence. The other is the grants that fund local media innovations.

The first requires the kind of rethinking about travel journalism discussed in Chapter 1, that is, the travel journalist as an independent foreign correspondent, traveling on behalf of substantive stories. *Editor & Publisher's* "2010 Journalism Awards and Fellowships Directory" lists 13 opportunities.[73] Some of them are place-based, such as the grants awarded by the East-West Center for work in Singapore, Taiwan, Hong Kong, and Korea. Others focus on investigative reporting, such as the grants from the Fund for Investigative Journalism. Some are linked to major universities, such as the IRP Fellowships in International Journalism with ties to the School of Advanced International Studies at Johns Hopkins University. Others are for extended periods of reporting, such as the Alicia Patterson Foundation Fellowships, which support six-month and 12-month reporting projects. Some are for visual journalists, such as the W. Eugene Smith Grant. Others are associated with global service organizations, such as the travel grants from Rotary International.

The second underutilized form of grant funding–local media innovations–is the newest and fastest-growing source of funds for media and journalism. Two factors underpin the growth. First is the decline in traditional outlets for journalism, brought on by disruption of the business models for newspapers and local television. Newsroom employment declined 33 percent—from more than 60,000 jobs in 1992 to about 40,000 in 2009, according to estimates by the American Society of News Editors.[74] Second is the emergence of digital platforms for journalism, and with it, a recognition that communities are not equally served. The result is a decline in the availability of local news. In a 2008 study, 41 percent of community leaders across the U.S. said the volume of local news was shrinking. Two years later, the number had grown to 75 percent.[75] Because of this, a wide range of journalism-oriented and place-based foundations have invested in media and journalism grants at the local, state, and national level.

Travel journalism might benefit from these grants. There are several reasons for this. First, travel and tourism are substantial—and largely under-covered—industries in many communities. Second, tourist destinations in many communities are arts, culture, and history institutions that donors to place-based foundations favor. And third, entrepreneurial travel journalists have demonstrated the innovativeness that journalism-oriented foundations want to support.

One area receiving a lot of grant-funding attention is public-service journalism, including explanatory, enterprise, and investigative reporting. One report indicates that grants fund all or part of 60 such news organizations, large and small, across the U.S.[76] How might travel journalists benefit from this? Nearly 60 percent of the SATW editors who responded to my survey somewhat or strongly agree that travel and tourism present an excellent opportunity for watchdog-type coverage.[77] The archives of Investigative Reporters and Editors contain many and varied examples of watchdog-type coverage of travel and tourism.[78] Yet, we know from other studies that far more can be done.[79] That is the opportunity.

# User Funding

User funding comes from two sources. One is directly from users—sometimes referred to as the "crowd"—in the form of payments such as donations, fees, or subscriptions. This source of user funding is typically for projects and start-ups. The other source is indirectly from advertisers, underwriters, or sponsors who pay, in effect, for access to users.

So-called "crowdfunding" has received the most attention among the various forms of direct user funding. Crowdfunding involves people contributing money, usually through the Internet, to support efforts of other people or organizations. Crowdfunding efforts support a variety of purposes, including travel journalism. Like grant funding, it's typically for one-time projects and start-ups.

Kickstarter is perhaps the best-known crowdfunding organization, and along with its competitor IndieGoGo, the most suitable for travel journalists. Launched in 2009, it raises and distributes about $1 million a week.[80] "Writing and publishing," "Food," and "Photography" are three (of 13) travel-related Kickstarter project categories.

Spot.us is, perhaps, the best-known of the journalism-specific crowdfunding organizations. A nonprofit, Spot.us was founded by David Cohn with funding from the Knight Foundation. Its best-known project was its first—freelance journalist Lindsey Hoshaw's work for the *New York Times* on the Great Pacific Garbage Patch, described by one critic as "good travel writing with a dash of garbage-related journalism thrown in."[81] Three years later, almost none of its projects focus on travel journalism.

An excellent tip sheet on crowdfunding is at socialmediaexaminer.com.

Advertising, underwriting, and sponsorships are indirect forms of user funding. A publication, whether print or online, delivers readers or viewers to advertisers who pay the publication to display the messages. These are common, well-understood[82] forms of funding travel blogs and websites owned by the freelance journalists I surveyed. Some 50 percent who responded rely on advertising, underwriting, or sponsorships. Advertising pushed to them by a network service such as Google's AdSense is simple to arrange, but pays little. Better to sell advertising, underwriting, and sponsorships directly, but you'll need a substantial audience. Matt Kepnes says he had 7,000 to 8,000 regular followers of his NomadicMatt site before advertisers expressed interest.[83]

The opportunity for advertising, underwriting, and sponsorships online should improve over time, according to an analysis Borrell Associates prepared for this text.[84] Borrell modeled local-market spending in six travel-related categories—amusement parks, gambling casinos, ground transportation, accommodation, tourist attractions, and travel services—across all media types. Borrell forecasts that between 2010 and 2016:

- Online travel-related advertising will increase 91 percent, from $218 million to $416 million
- Online travel-related promotions spending will increase 277 percent, from $13 million to $49 million

By contrast, Borrell forecasts that travel-related advertising in newspapers and other print, which includes magazines, will decline 12 percent, from $473 million to $414 million.

## Investor Funding

Investor funding is last on the list for good reason: It's the least likely to be successful. It's on the list, however, because a number of travel-related start-ups have been funded by investors. The earliest stage of investor funding is known as "seed funding." It often comes from friends and families. Next is "angel funding," in which the funds typically come from a wealthy individual who invests in return for equity shares. Angel investors pumped $20 billion into start-ups in 2010, up 13.6 percent over 2009.[85] Then there are the "rounds" of funding by venture capital funds: The usual progression is from start-up to first round to mezzanine to pre-IPO. Several travel-related examples illustrate the process. An entrepreneur named Thomas Owadenko launched an online video hotel guide called Trivop in early 2007. Later the same year, he raised $800,000 from seven "angel" investors, plumping Trivop's aim of "becoming the de facto Internet videoguide for the hotel industry."[86] Reviewers referred to Trivop as "the traveler's YouTube."[87] But Trivop wasn't alone in the hotel video space. TVTrip was there, too, backed by $4.8 million from a pair of "early-round" venture capitalists.[88] So was Travelistic.com and, to a lesser extent, LonelyPlanet.tv.[89] As Trivop grew, its cache of hotel videos became attractive to the hotel reservation service hrs.com, which acquired Trivop in 2009 for an undisclosed sum.[90] Investor capital, hard as it is to obtain, does not assure success: Travelistic.com is out of business. Nor is seeking investor capital for the timid or the naïve. Thorny, hard-to-understand regulatory issues stand in the way.

# Some Other Nontraditional Sources of Funding for Travel Journalism

Study abroad as a student travel journalist—Study abroad is the best, first opportunity to explore travel journalism. At least 10 U.S. universities offer study abroad programs in travel journalism. Some programs target traditional study abroad venues such as Italy, Australia, and Costa Rica.[91] Others traverse Asian countries such as China. My school, the University of Georgia, offers a program based in Cambodia.[92] Most focus on reporting and writing, some on photography. All offer publishing opportunities. Study abroad is a relatively low-cost method to fund a first opportunity to explore travel journalism. And many offer scholarships, especially to honors students traveling to nontraditional study abroad venues.[93] Moreover, whether the focus is travel journalism or another topic, research indicates that study abroad improves academic performance, graduation rates, and knowledge of cultural practices and context.[94] One limitation: Some of the most journalistically interesting study abroad venues involve risk, and study abroad programs tend to be risk averse. In 2010, for example, Northeastern University cancelled study abroad programs in four countries: Thailand because of antigovernment unrest, New Zealand because of earthquakes, Egypt because of revolution, and Japan because of the earthquake, tsunami, and nuclear crisis.[95]

Teaching English as a Second Language abroad as a sinecure—Lillie Marshall is an exemplar of teaching English abroad as a way to fund travel journalism. One summer, the high school English teacher took the class that certified her as a teacher of English as a Second Language. Then she left her Boston classroom for a nine-month world journey—mostly travel, some teaching—which she blogged about at aroundtheworld.com. Today, she is back in the classroom in Boston, traveling on breaks, writing about travel—and encouraging other teachers to do the same. "The point of TeachingTraveling. com," she writes on her new blog, "is to celebrate that teaching is great, traveling is great, and we should all do more of both. This site aims to inspire more teachers to travel and more travelers to teach."[96] Teaching English as a Second Language is one smart way to start. Brief, immersive assignments in foreign classrooms are, in themselves, excellent travel journalism venues. Moreover, they fund the travel to there and elsewhere—a sinecure. Developing countries offer the most opportunities. Training and certification are open issues. "Training is optional," Marshall says, "but useful."[97] Avoid online-only courses. They lack the hands-on experience vital to teaching. Look instead for the in-classroom courses that lead to certification by Cambridge University's Certificate in English Language Teaching to Adults.[98] Find them in 54 countries, Cambridge says.

Take a "career break" to travel and write—Nomadic Matt's Matthew Kepnes launched his career as a travel journalist through the increasingly popular, moderate-risk practice of the "career break." Kepnes was an MBA-trained hospital administrator in Boston who decided to travel and write for a year—and never went back to the cubicle.[99] For most people, career breaks—sometimes called sabbaticals—aren't transitions from one thing to another, as happened with Kepnes, but planned, temporary—often strategic—interruptions in work, often with the employer's approval and support. Some call it the "Eat, Pray, Love Effect," after Elizabeth Gilbert's best-selling memoir.[100] "The popularity of 'Eat, Pray, Love' and the tremendous proliferation of travel blogs," says Tara Russell, "shows that we're fascinated by the idea of escaping our everyday lives."[101] Russell is a leader in the emerging career-break infrastructure. She is a career-break coach in San Francisco,[102] allied with a group that conducts "meet-plan-go" workshops across the U.S.,[103] and others who manage a travel blog about career-break experiences, briefcasetobackpack.com. Each offers advice about negotiating with employers, planning, and financing career breaks.

Embed on a mission, an interest group "fact-finding" trip, a military operation, or an alumni organization tour—The concept of journalists "embedding" became common during the U.S.-led conflict in Iraq. Embedding was a way that journalists could cover operating forces—usually combat forces—at a cost and level of safety superior to going it alone. Though controversial, nearly 800 journalists embedded with U.S. troops at the start of the war.[104] It's a concept, too, for funding travel journalism, if the risks and cautions are well understood. There are many opportunities in most communities: Embed with a religious, medical, or humanitarian mission; follow an interest group on a "fact-finding" trip; embed with troops deploying to an area of interest; join a college alumni organization tour. The journalistic opportunity is two-fold. First, it is to cover the mission, "fact-finding" trip, deployment, or tour, and second, it is to develop independent stories at the place or places you visit. Either or both can be travel-related. Be aware of the risks and cautions. Journalists who embedded with the U.S. military were criticized for relinquishing independence, accepting restrictions, and appearing too sympathetic.[105] What ground rules might travel journalists set to avoid the same criticisms? Pay a fair share of the expenses. Accept no restrictions on movement or coverage. Disclose all in the coverage.

## Multiple Sources of Funding Are a Best Practice

No one source of funding—traditional or nontraditional—is the silver bullet. Indeed, the best approach is developing multiple sources. My survey of freelance members of the Society of American Travel Writers indicates that those who own travel blogs and websites tend to rely on one or more of three sources of funding—advertising, affiliate marketing and direct sales, and user funding. Among those who responded to my survey:

- 58 percent rely on one or more of the three sources
- 42 percent are self-funded with no other apparent or disclosed source of funding
- 50 percent rely on advertising or "sponsorships," either arranged themselves or pushed to them by an advertising network service such as Google's AdSense
- 29 percent rely on product sales, usually guidebooks or travel gear, typically through marketing arrangements with affiliate companies, or directly through the blog or website
- 4 percent rely on voluntary payments from users

Only one site relies on all three. It is Roads Traveled, a regional site focusing on Wisconsin, owned by long-time print travel journalist Mary Bergin.[106] The site features five locally sold ad positions as well as a network-fed AdChoice position. There is a product sales channel as well as a donation channel. But Bergin would be the first to tell you that multiple revenue streams do not guarantee business success. In August 2011 we found ourselves together in a Poynter seminar on entrepreneurial journalism, struggling to figure out how to make sites such as Roads Traveled pay the bills.

Another best practice is to travel on the cheap. Tips on how to do so are discussed in Resources.

## A Caution about These Approaches to Funding Travel Journalism

One of the freelancers I surveyed suggested this method of funding travel journalism: "Marry a wealthy . . ." Fine, of course. It's also fine to be an all-expenses-paid employee of a top newspaper Travel section, a glossy travel magazine, or a guidebook. But for the rest of us, it's one or more of the nontraditional sources noted here. Each involves risk. One is the loss of independence, as the travel journalist comes to depend on third parties for support. Advertisers may want links to their sites embedded in editorial content. Grant makers may want editorial content to focus on places or themes consistent with their goals. Investors may want . . . and so forth. It's easy enough, in the abstract, to "just say no," but such pressures are a regular constraint. Two keys: One is to negotiate, in or out of an agreement, any such understandings. The other is to disclose all such understandings to readers.

## Test Your Understanding

1. Describe how "media tours" work. Why is conflict of interest embedded in the process?
2. Travel journalists are not of one mind with respect to accepting and disclosing subsidized travel. Describe the different arguments they frame.
3. Given these arguments, what is your position on accepting and disclosing subsidized travel?
4. Three ideas emerged from the research with respect to codifying disclosure of free and reduced-rate travel. Which one do you regard as the most important and likely to be adopted?
5. This chapter identifies four nontraditional sources of funding travel journalism. Which one best applies to you and why?
6. Among the "other" nontraditional sources of funding travel journalism is teaching English as a second language abroad. Does this appeal to you and why?

## Practice Your Skills

Assignment #7—For the story or website outlined in earlier assignments, draft a funding proposal at a user-funding site such as Kickstarter. If you use Kickstarter, go to http://www.kickstarter.com for guidelines, requirements, and a template to submit your proposal. It's fine to work offline as an exercise. But why not try it for real? The Kickstarter staff will provide feedback, as will the user-funding marketplace.

## A Closer Look: Crowd Funding a Travel Book about Elvis Presley's Birthplace

When Jane Blunschi was looking for $2,500 to support a travel book about Tupelo, Mississippi, her publisher suggested she try Kickstarter.

Kickstarter is a crowdfunding application that focuses on creative projects.

So Blunschi put together the elements of a Kickstarter appeal: a video, a blurb, a list of rewards for contributions, and a short bio.

"The world NEEDS to know about Tupelo, Mississippi," Blunschi wrote. "Love rock and roll? Blues, gospel, country music? Thank Tupelo. No Tupelo, no Elvis Presley . . . know what I mean?"[135]

Two months later, 30 backers contributed the $2,500, and Blunschi was good to go.

Blunschi's experience is a good example of how a writer uses crowdfunding to finance a project.

Crowdfunding entails people contributing money, usually through the Internet, to support efforts of other people or organizations. Crowdfunding efforts support a variety of purposes, including travel journalism. Kickstarter is, perhaps, the best-known crowdfunding organization.

"Kickstarter is the largest funding platform for creative projects in the world," according to its site. "Every week, tens of thousands of amazing people pledge millions of dollars to projects from the worlds of music, film, art, technology, design, food, publishing, and other creative fields."[136] Kickstarter funds projects such as performances, films, writing. It does not fund charities or causes.[137]

Travel journalists are among the applicants drawn to Kickstarter.

Launched in 2009, it raises and distributes about $1 million a week. "Writing and publishing," "Food." and "Photography" are three (of 13) travel-related Kickstarter project categories.

Filmmaker Pegi Vail raised $5,958 to complete a documentary about backpacker tourism, "Gringo Trails."[138] Thirty-five backers gave Kristen Schwarz a total of $1,528 to attend a writing conference in Positano, Italy, where she planned to "meditate on . . . the nature of travel writing."[139] Student journalists at the University of Alabama raised $3,670 to "report, shoot, edit and design a 140-page [travel] magazine," *Alpine Living.*[140] The students used the money to travel to Switzerland and Germany.

Kickstarter is just one of a dozen crowdfunding applications. IndieGoGo, like Kickstarter, focuses on creative projects. Buxxbnk and 33needs target social innovations. Photojournalists turn to emphas.is. Nonprofits fundraise at FirstGiving or CauseVox. ProFounder underwrites traditional ventures. AppBackr, perhaps the most niche-like, focuses on mobile app enterprises.

I caught up with Blunschi by email just as the Tupelo project was funded.

Blunschi was introduced to Kickstarter by her publisher at Corvus Press, Jeremy Broussard. "I was into the somewhat structured approach to fundraising," Blunschi told me. "I had never before heard of Kickstarter, but I have told a ton of people about it since I launched the project."[141]

The process is entirely online. The applicant registers with Kickstarter, answers a few questions about eligibility, then uploads material about the project, typically a video, a text blurb, rewards offered to backers, a bio, and a link to the project website. I wanted to know if there was any person-to-person interaction, such as advice about the video or about the reward offers.

"Not a peep," Blunschi wrote. "A-okay with me."

How about the pros and cons of using Kickstarter, at least in that early part of the process?

"Pros—The site, especially my video, was a lovely, concrete 'calling card' for my book. I cannot tell you how many of my friends, family, and business associates copied the link to Facebook, and that really served to start a buzz on the project. Plenty of people who did not pledge let me know that they had seen the video and encouraged my effort, and that let me know that the word was getting out there. Also, it saved me from having to beat the bushes as far as asking folks for contributions. I was able to connect the essence of my project with my personality and creative influences in my Kickstarter video. I *wanted* people to see my grandma's Elvis Presley concert tickets, and know some of the reasons I am creating this book. I wanted them to hear me, in my own (heavily accented) words say how much the opportunity to make this book means to me."

"Cons—I have spent a lot of time breaking down the ins and outs of Kickstarter and reassuring people that pledging online is simple and secure. However, I think that's just a symptom of growing pains; I expect Kickstarter to go off like a rocket all over pretty soon."

Brunschi says she would have done a few things differently. She would have timed the project much closer to the release date of the book *Love, Tupelo*, and mentioned the rewards in the video.

Any advice for others?

"Kickstarter videos and project sites are extremely addictive," she wrote. "This is a pro and a con. Whole afternoons—lost. Not kidding."

The book was published in the fall of 2011.

# AFTERWORD

As with many things, this text grew out of a need.

I was preparing to teach a study abroad program in travel journalism. None of the dominant texts in the field was right. Most focused on how to write—and sell—travel experiences. None was grounded in the journalistic identity, purpose, and method we emphasize to our students.

Indeed, the word *journalism* rarely, if ever, appeared in these "how-to" texts.

This seemed off to me. I wanted the students to approach travel much as they would approach any other beat. Whatever the type of story—News, Service and Advice, Destination, or Journey—it was to be reported and written to a journalistic standard.

As class began on a hot spring morning in Siem Reap, Cambodia, I remember telling the students that 80 percent of travel writing is not journalistic. It lacks the independence, ethics, and substance characteristic of journalism.

"We're going to model the other 20 percent," I told them.

And from that need the idea for this text emerged. I began to test with these students an argument I'd been toying with for some years. As framed in the Introduction, it is this:

Travel writing and travel journalism are not the same. Travel writers and travel journalists are not the same. They differ according to identity, purpose, and method. The travel writer looks in a mirror, tending toward memoir and autobiography. The travel journalist looks through a window—what's outside is what matters. The travel writer depends on and serves the travel industry. The travel journalist is independent of the travel industry and serves the public. The travel writer is subsidized. The travel journalist pays her own way.

I came to this argument over time. My own work was a starting point. Over the last 20 years I've reported from 25 countries on five continents. The work moved from lightweight fare to stories of real substance. The first piece took me to Nick Bollettieri's tennis academy in Bradenton, Florida. It contained lines such as "it was a week of combining backhands and suntans." The climatic moment was a test match the last day with Anna Maria Foldenyi, an 11-year-old phenom from Hungary. I lost, badly.

More recently, I examined post-apartheid inequality in South Africa. Set in Johannesburg, my piece contrasted life in suburban Sandton, perhaps the richest square mile in South Africa, with the nearby township of Alexandra, certainly the poorest. A main character was a 5-year-old boy named Gif, who was paying for school by selling puffed snacks at a shop his family operates on a corner. It's one of many informal businesses in Alexandra—a few tables shielded from the sun by canvas tenting.

The story concludes: "As I left the region, I wondered what will happen to Moshidi's son, Gif. Like Solomon, will he get out? Like Sello, become a doctor? Like Abbey, stay and build a business? Or, like too many of Solomon's childhood friends—end up dead?"

Struggling school boys like Gif, more than tennis prodigies like Anna Maria, were on my mind as I started to work with the study abroad students in Cambodia. With little more than direction and encouragement, could these students find the Gifs of Cambodia? Could they find and introduce characters through which readers might come to understand broad, substantive stories? Could they apply all of the sourcing tools available to journalists, not just observation, so overused in travel writing? Could they overcome language barriers, working through translators? Could they do all of this in 15 days?

The work they produced—as described in case study #1, "Students Test a More Journalistic Approach to Travel Writing"—strengthened my confidence in this text's central argument. Even relative beginners, parachuted into a strange culture, can produce travel narratives to a high journalistic standard: independent, ethical, substantive.

I'm less confident, however, about the business model I propose for travel journalism. The entrepreneurial approach described in Chapter 6 and the nontraditional funding described in Chapter 7 are largely untested. There is substantial, anecdotal evidence that travel journalists are succeeding with both. But there is no systematic evidence—yet.

This mix of confidence is fine with me. Everything in journalism is up for rethinking. Why not travel writing? Now is the time for new ideas, fast prototyping, test marketing, and proof of concepts.

Let me hear from you about your experiences—what works, what doesn't.

John F. Greenman
University of Georgia
January 4, 2012

# RESOURCES

What follows are resources tailored to the travel journalist. No claim is made about comprehensiveness. Rather, these are selections based on credibility, access, and cost.

## Organizations

Organizations of interest to travel journalists fall into five categories. First are the professional organizations that limit membership to established journalists. Second are the community-like organizations that link travel journalists with each other. Third are the exchange-like organizations that link travel journalists with travel marketers. Fourth are the niche topic–oriented organizations. And fifth are local writing clubs.

## Professional Organizations

Three professional organizations stand out for providing meaningful benefits at a reasonable cost. They are the Society of American Travel Writers (SATW), the American Society of Journalists and Authors (ASJA) and the National Writers Union (NWU).

None offers membership to beginners. All require evidence of published work—or, in the case of the NWU, finished work an applicant is attempting to publish. Each has the kind of infrastructure—offices, staff, boards of directories, publications, annual meetings—that demonstrates stability and effectiveness over time.

Here's a bit on each:

Founded in 1956, the Society of American Travel Writers is among the oldest and most prestigious organizations of travel journalists. Membership blends travel editors and freelance writers and photographers, 57 percent, with travel industry publicists, 43 percent. Standards to join, especially on the editor, writer, photographer side, are high: lots of published work plus sponsorship by two SATW members. Like many of journalism's professional organizations, SATW is under pressure from industry decline. Member benefits include networking, workshops, publications, legal consultation, travel discounts. Note: Travel journalists do not need to be a member of SATW in order to enter the Lowell Thomas Awards competition. Contact information: www.satw.org

The membership of the American Society of Journalists and Authors is both broader and narrower than SATW: It is broader in that it embraces all genres of nonfiction writing, not just travel; it is narrower in that its membership is limited to freelance writers—no editors, no publicists. Like SATW, standards to join are high: again, lots of published work. But sponsorship is not required. ASJA positions membership as a "credential" with the benefit of a support group. Benefits include insider access to data about markets, contracts, rates, trends. It also offers member-only meetings and awards. Contact information: www.asja.org

The National Writers Union positions itself as "the only labor union that represents freelance writers."[1] NWU is Local 1981 of the United Auto Workers. Its members are freelance writers in any genre, and as freelancers they do not benefit from a collective bargaining agreement, as do employed journalists represented by an American Newspaper Guild local. Nonetheless, NWU offers union-like benefits such as contract advice and grievance assistance. The press pass offered to NWU members may be its most important benefit, according to veteran freelance writers: "This laminated photo ID . . . will prove adequate in virtually all circumstances."[2] Contact information: www.nwu.org

Some professional organizations similar in membership requirements and benefits to SATW, ASJA, and NWU are focused on specific regions of the U.S.

They include the Midwest Travel Writers Association, the New York Travel Writers Association, and the Bay Area Travel Writers.

Canada has two professional organizations, the Professional Writers Association of Canada and the Travel Media Association of Canada.

## Community-like Organizations

Travelblogexchange.com and travelwritersexchange.com are examples of community-like organizations of travel journalists—especially those working online. Both are commercial sites with sponsors, and their memberships and subscriptions are free. Both exploit the power of social media to connect their members through forums, discussion groups, and—in the case of travelblogexchange.com—meet-ups and conferences. Both offer the opportunity to showcase work and get and give advice.

## Exchange-like Organizations

Travelwriters.com is probably the best example of an exchange-like organization linking travel journalists with travel marketers. "Travelwriters.com is based on a simple principle," the website asserts, "to connect top-tier writers with editors, PR agencies, tourism professionals, CVBs and tour operators, nurturing the important link that so heavily influences the travel media." Founded in 1997, travelwriters.com boosts 1,000 members. Membership is free to writers and editors. The PR agencies, CVBs, and tour operators pay per year or per event to advertise press trips and media tours. Contact information: www.travelwriters.com.

## Niche topic–oriented Organizations

Probably the best-known niche topic–oriented organization is the International Food, Wine and Travel Writers Association. The chief member benefit is the subsidized travel—focusing on food and wine. Three to five trips are offered yearly. A nominal registration fee, typically $150, covers air travel, hotels, food, and wine for a six-day trip. In return, members are expected to submit evidence to their hosts that they wrote and published something about what they saw and tasted. The admonition, "No clips, no future trips," is printed in bold on the registration form. Members must pledge to show up on time, dress appropriately, not order alcohol at lunch, and listen when the host is talking. Contact information: www.ifwtwa.com. Another food-related organization that includes writers and photographers among its member is the International Association of Culinary Professionals. Other niche-topic organizations focus on cruise, adventure, and sport travel.

## Local Writing Clubs

Local writing clubs vary in size and sophistication, but all share one goal: to bring together writers who live in a community or region to discuss and help each other with their work.

Writer Andrew Hempstead says the best way to find a local writing club is to search online for "(your city) writing club."

Don't expect much infrastructure, especially in smaller markets. But that's a strength, in a way. Local writing clubs should be small, highly personal support groups. Writers should be coming together to share their writing, and to seek advice about problems with their writing.

Also, don't expect to find a local writing club that focuses just on travel writing, except in the largest markets.

## Seven Essential Readings for Travel Journalists

More than 15,000 titles pop up when you search "travel writing" at Amazon. com. Too many to skim, let alone read. Here's a much shorter list—just seven titles—that every travel journalist ought to read.

## If You Only Have Time for One Travel Writing "How-to" Text

A half-dozen texts dominate the "how-to" market.[3] It's fine to read them all, though much of the advice is the same in all of them. The travel journalist with time to read just one should order *The Travel Writer's Handbook: How to Write—and Sell—Your Own Travel Experiences*. Louise Purwin Zobel and Jacqueline Harmon Butler have sustained this title over six editions—more than two decades—by providing comprehensive, coherent, accessible advice. Both Zobel and Butler are experienced travel journalists, and tales from the road contribute verisimilitude. The chapter "Writing for Others to Read" demonstrates a welcome healthy respect for the audience.

## Hearing the Voice of Other Travel Journalists

A number of excellent texts contain interviews or how-I-did-it essays with travel journalists.[4] The best is *They Went: The Art and Craft of Travel Writing* (1991), edited by William Zinsser. This is the sixth in a series of books about writing edited by Zinsser, and his own insights about travel journalism are a key contribution. Another strength is the advice about sense of place—a central characteristic of travel journalism—from the writers whose essays Zinsser includes. Though most of the writers are known for their book-length narratives, their advice is applicable to travel journalism pieces of any length.

## Understanding Travel Marketers and Their Marketing Practices

Travel marketers influence travel journalism in myriad ways. It's important, therefore, for the travel journalist to understand travel marketers and their marketing practices. A widely used travel marketing text is *Destination Marketing* (1988) by Richard B. Gartrell. Gartrell's background is travel marketing, and he brings an industry sensibility, not an academic tone, to this text. Well-thumbed copies are likely to be found in every travel marketer's library. For an academic take, consult Steven Pike's *Destination Marketing: An Integrated Marketing Communication Approach* (2008), and its companion text, *Destination Marketing Organisations* (2005).

## The Best One Inch of Reading about Entrepreneurial Journalism

*Journalism Next* (2010) by Mark Briggs is an indispensable "how-to" book for the entrepreneurial journalist. Dan Gillmor describes it as "high-level concepts plus nuts-and-bolts specifics on how to proceed, in a creative and engaging format."[5] We've come to expect this approach from Briggs, whose *Journalism 2.0* (2007) made the case that better journalism will evolve from leveraging digital technology. How best to use this book? Begin with an idea for a travel-related entrepreneurial project—say, a travel blog about a local-market destination. Apply the idea to the 29 "What's Next" exercises. By the end, you'll have traveled the journey from RSS feeds to blogs, audience engagement, audio, video, and monitoring. And you'll have your travel-related idea fully in the marketplace. A related opportunity is Briggs's online course for Poynter's NewsU, Becoming an Entrepreneurial Journalist: From Idea to Implementation.

# Writing Primers: Ideas, Drafting, and Revision

Ideas—James B. Stewart's *Follow the Story: How to Write Successful Nonfiction* is the outgrowth of his work as Page One editor of the *Wall Street Journal*. The first 45 pages are devoted to developing story ideas. "A good idea constitutes about 50 percent of what makes a successful story," Stewart writes.[6] Then he describes a process for finding and developing ideas.

Drafting—Anne Lamott's *Bird by Bird: Some Instructions on Writing and Life* deals smartly with a common affliction among writers: writing the first draft. "All good writers write . . . shitty first drafts," Lamott writes. "That's how they end up with good second drafts and terrific third drafts."[7] Lamott's wisdom on drafting consumes just seven pages. It's worth reading before drafting—every time.

Revision—Roy Peter Clark's *Writing Tools: 50 Essential Strategies for Every Writer* does, in fact, contain 50 writing "tools," each one a "how-to" with concepts and exercises. Practice one tool each week. After a year, you're a better writer. Repeat. Repeat again. I encourage my students to use the tools for revision rather than drafting. As noted in Chapter 4, I teach Tools #1–10 under "Nuts and Bolts" to my writing students. Over time, of course, applying the tools in revision improves drafting.

# Academic Experts

Dozens of U.S. universities (and more across the world) fund programs, centers, or institutes in travel, tourism, and hospitality studies. They offer a broad range of services: certificates and degree programs, research and consulting, and public service and outreach. For travel journalists, they offer reportable research and access to the academic experts behind it.

Some examples of how this might work:

Local-market reporting—A story that examines the economic impact of a proposed event—say, a sporting event some market is competing to host—would benefit from an academic expert's research. Every region of the U.S. and many states employ academicians whose research focuses on measuring the economic impact of events.

Niche-topic reporting—Academicians focus on tourism niches just as the travel industry and travel journalists do. The best way to find them: Scan "research interests" under faculty profiles at universities with travel, tourism, and hospitality programs. My scan at just one university turned up faculty with research interests in four niche topics: cultural, heritage, environment, and sports.

Policy-issues reporting—Policy disputes emerge regularly in this large and highly regulated industry. No policy issue is broader or more debated than "sustainability." The Center for Sustainable Destinations lists 10 international policy organizations and 10 academic institutions focused on travel and tourism sustainability policy.[8]

An exhaustive list of programs, centers, and institutes is beyond the scope and purpose of this text. Indeed, the list is so long and varied that, no single simple search will reveal the offerings; a better approach is to consult academic journals. Nevertheless, here are three examples that demonstrate the range of offerings.

- Some institutes offer a global reach. One of the best known is George Washington University's International Institute of Tourism Studies, which has been involved with tourism development and education for over 20 years. The Institute's faculty are expert in training and development, technical assistance, assessments, and consulting. They work locally to globally in close alignment with the United Nations's World Tourism Organization. Topics of interest include tourism destination management, event management, hospitality management, conservation, and parks and protected areas. The Institute is one part of a broader academic program within the Department of Tourism and Hospitality Management that offers certificates, undergraduate and graduate degrees in sustainable tourism destination management, sports management, events management, and hospitality management. The Institute's director is Dr. Kristin Lamoureux.

- Some institutes focus on a state or region. The Center for Tourism Policy Studies at the University of Hawaii, for example, focuses on the state of Hawaii as well as countries in the Asia-Pacific region. The Center conducts research for academic, government, and industry use, and offers professional development and technical assistance. Like the institute at George Washington University, the Center is part of a broader academic program—it falls under the University of Hawaii's School of Travel Industry Management, which offers undergraduate and graduate degrees. The Center's director is Dr. George Ikeda.
- Some institutes focus on a policy interest. As its name implies, East Carolina University's Center for Sustainable Tourism focuses on sustainability. It claims that it is the only center with this focus in the U.S.[9] It researches sustainable practices in the travel and tourism industry, advocates for policies that promote sustainability, and prepares students for management, research, and teaching careers. The Center's director is Dr. Patrick T. Long.

## Industry Experts

The U.S. Travel Association is the best single source of industry expertise. Its website, www.ustravel.org, is the starting point. Three of the channels are particularly useful. The Research channel contains industry data about domestic and international travel as well as the economic impact of travel. The Government Relations channel tracks U.S. Travel Association lobbying, mostly at the federal level. The News Channel contains an archive of travel industry coverage—a good barometer of what's newsworthy about travel. The association's staff is member-focused, but it has a press office.

## Trade Publications

*Travel Weekly*, now 54 years old, is the starting point for any travel journalist's trade publication library. It bills itself as "the national newspaper of the travel industry," and its content includes a lot of day-by-day travel industry news coverage. But of equal value are special issues devoted to, for example, travel consumer trends. *Travel Weekly* is owned by Northstar Travel Media, which publishes 15 titles and 10 newsletters, all online and all free.

By no means is *Travel Weekly* the only trade publication for the travel industry. *Bacon's Magazine Directory* lists 50 travel trade publications that range in focus from bus tours to travel agents and destinations.[10]

# Lowell Thomas Award Recipients, 2000–2011

Travel journalists, like all journalists, tilt toward outlets that other journalists admire: Peer approval is journalism's gold standard. Among travel journalists, the gold standard is the Lowell Thomas Awards, conferred annually by the Society of American Travel Writers Foundation.

The Lowell Thomas Awards, which began in 1985, often are referred to as travel journalism's Pulitzer Prizes. Like the Pulitzers, the Lowell Thomas Awards are administered by a school of journalism, currently the University of North Carolina at Chapel Hill. Awards are made in 25 categories in print and online. Lowell Thomas Awards are taken seriously by the winners: After its first gold award in the category Best Travel Magazine, *Afar* redesigned its cover to show off its win: "Winner: America's best travel magazine."

## Newspapers

Each year the Lowell Thomas Awards honor newspaper Travel sections in four levels of accomplishment—gold, silver, bronze, and honorable mention—according to circulation size. Using this award as a proxy for "quality," I determined which newspaper Travel sections have been honored the most frequently over the last decade—2000 to 2010.

They are, in order:

- *New Orleans Times Picayune*, nine awards
- *Boston Globe*, eight
- *Orange County Register*, seven
- *Ottawa Citizen*, six
- *Chicago Tribune*, six
- *Los Angeles Times*, six
- *San Jose Mercury News*, five
- *Seattle Times*, four
- *Atlanta Journal Constitution*, four
- *Baltimore Sun*, four

# Magazines

The Lowell Thomas Awards also honor travel magazines in the same four levels of accomplishment. Those honored the most frequently over the 10-year period (2000–2010) are:

- *National Geographic Traveler*, eight awards
- *Budget Travel*, six
- *National Geographic Adventure*, five
- *Travel + Leisure*, four
- *Arizona Highways*, two
- *Via*, two
- *Caribbean Travel & Life*, one
- *Virtuoso Life*, one
- *Afar*, one

# Websites

The Lowell Thomas Awards also honor travel websites in the same four levels of accomplishment. Those honored the most frequently over the 10 years are:

- lonelyplanet.com, five awards
- nationalgeographic.com/traveler, four
- worldhum.com, three
- newyorkology.com, three
- budgettravel.com, three
- cruisecritic.com, two
- boston.com/travel, two
- frommers.com, two
- reidguides.com, one
- turkeytravelplanner.com, one
- wildwritingwomen.com, one
- alaska.com/akcom/travel, one
- southernliving.com/southernbyways, one
- matadornetwork.com, one

There are other awards for travel journalism, but none as prestigious as the Lowell Thomas Awards. Their results are similar—the same names crop up on their awards lists. Entrants for the Lowell Thomas Awards also compete for awards from the North American Travel Journalist Association Awards. Magazine entrants also compete for *Folio's* Eddie Awards. Website entrants also compete for Webbys and Bloggies. And so forth.

# Where to Learn More about Entrepreneurial Journalism

As noted in Chapter 6, journalists confront several disadvantages when it comes to entrepreneurship. Help for them ranges, on a continuum of least to greatest investment of time, from books, online courses, and boot camp–like seminars and workshops, to undergraduate classes, masters degree programs, and mid-career fellowships.

Here are some examples:

Journals—In the online "Professor's Corner" the Nieman Foundation has archived stories, reviews, and advice about "Entrepreneurial News Reporting: Digital Approaches" that were first published in *Nieman Reports*.[11] The impetus was to provide journalism school faculty with teaching tools, but the materials are accessible to any journalist seeking instruction.

Books—*Journalism Next* by Mark Briggs is the best book about entrepreneurial journalism. See above under "Seven Essential Readings for Travel Journalists."

Online courses—Poynter's NewU's offers two self-directed courses and a regular diet of webinars for entrepreneurial journalists. One of the courses, Mark Briggs's Becoming an Entrepreneurial Journalist: From Idea to Implementation, reinforces many of the concepts and skills in his book *Journalism Next*. Another course of interest to innovators is Targeting New Audiences: Finding Your Niche. Webinars include Innovation on the Front Lines of Journalism and Secrets of Successful News Startups.

Boot camp–like seminars and workshops—The Knight Digital Media Center has been holding annual three-day workshops in Los Angeles since 2009. A week-long multimedia boot camp is offered by the Freedom Forum in Nashville. Another week-long seminar, Digital Entrepreneurship, is offered by Poynter in St. Petersburg. We Media holds a one-day boot camp for "would-be entrepreneurs" as a part of its annual conference.

Blogs to follow—PaidContent.org covers the business of digital media. Journalist Rafat Ali founded the blog in 2002 as a chronicle of the economic evolution of digital content. Ali sold PaidContent.org in 2010 to Guardian News and Media Limited, which has added other sites that report on digital business. Media Shift Idea Lab is a group blog by journalism innovators, funded by the Knight News Challenge. They explain their projects, share intelligence, and interact with the new-media community online.

Undergraduate classes—Arizona State University teaches more than 80 courses in entrepreneurship across 10 of its schools and colleges—including its Walter Cronkite School of Journalism. The ASU idea is to teach entrepreneurship as a tool to problem-solving. Entrepreneurship is not housed in a single college or institute; instead, it's "suffuse[d] . . . into the fabric of the university."[12] The journalism school's Knight Center for Digital Media Entrepreneurship offers one course, Digital Media Entrepreneurship, that is open to all ASU students. A detailed syllabus is available to the public.[13] ASU is one of 19 U.S. universities whose entrepreneurship programs are funded in part by the E. M. Kauffman Foundation.

Mid-career fellowships—Some mid-career journalism fellowships are reorienting their focus toward entrepreneurial journalism. One of them is the John S. Knight Journalism Fellowships at Stanford University. Gone is the idea of a one-year sabbatical, replaced by hands-on projects that focus on innovation, entrepreneurship, and leadership. Among the first fellows under the new focus is Burt Herman, cofounder of the blog *Storify*, which won the $10,000 grand prize in the 2011 Knight-Batten Awards for Innovations in Journalism.

Masters degree programs—The Tow-Knight Center for Entrepreneurial Journalism at the City University of New York offers a master of arts degree in entrepreneurial journalism. "What Stanford and MIT do for technology," says founder Jeff Jarvis, "we hope we can do for journalism." The program blends courses in general business and the business model for news with start-up–oriented courses and an apprenticeship. A course of shorter duration is the one-semester certificate program in entrepreneurial journalism. Syllabi for all courses are available to the public.[14]

# Travel on a Shoestring—Advice from Three Sources

Advice about traveling on a shoestring is everywhere. Book air travel on Tuesdays because airlines announce sale fares on Monday nights. Use yapta.com to track your fare after you've booked it, and request a refund if the fare drops.

Don't park at the airport. Don't pay to check a bag.[15] And that's just the advice about air travel.

No point in duplicating the well-trod land of travel tip sheets here. But there may be value in recent advice that is specific to the needs of travel journalists. Here is advice from three sources: a college student getting on the road for his first travel journalism assignment, 100 freelance members of the Society of American Travel Writers, and a range of sources focused on one topic—accommodation.

# Aaron Marshburn's "Making Friends Model"

The college student is Aaron Marshburn, who traveled and wrote during the summers of his undergraduate years at the University of Georgia's Grady College of Journalism. I supervised two of Marshburn's trips, one to Southeast Asia, the other to South Africa.

It was on the Southeast Asia trip when Marshburn evolved his model for inexpensive but effective travel. He calls this method the "Making Friends Model." In a summary he prepared for this text,[16] Marshburn wrote: "Some of the most telling details and most memorable characters are often found in the unlikeliest places, and they are easiest to find when you let yourself drift in and out of the lives of the people you meet on the road. This method is not one for those who need structure or fear the unknown, but rather for those who have an insatiable desire to experience each moment for its own sake and then take the time to write about it."

Here are some of his ideas, in tip-sheet form:

- Hitchhiking
- Banding together with groups of people you meet
- Hanging around in bars and restaurants
- Asking questions relentlessly, and never missing an opportunity to do so
- Consulting hotel/hostel staff, taxi drivers, locals you come in contact with
- Traveling through a web of friends of friends of friends
- Volunteering for temporary stints through NGOs, schools, TEFL
- Helping people—ANYONE!
- Wandering, exploring, getting lost, getting scared
- Learning toasts, greetings, etc. in a foreign language

- Approaching strangers, and taking a moment to listen to anyone who wants to speak with you—even beggars!
- Using www.couchsurfing.com
- When in doubt, just do it! Don't let key opportunities pass you by because you made too many fixed plans, or feel uncomfortable with something
- Being generous with your time, laughs, smiles, and sometimes money
- Buying local clothes—blend in. Don't be cleaner than everyone else
- Traveling overland; avoiding flights if possible
- Stopping anytime you see something out of the ordinary and finding out what's up

# Advice from 100 Freelance Travel Journalists

Freelance travel writers tend to show a sarcastic, even cynical side when asked for advice about money-saving travel tips.

"Marry rich" is one theme. "Keep your day job" is another. "Everything comp[ed]" is a third. "I have no idea what you're seeking here," one freelancer wrote in reply to my request for money-saving tips. "Stay at home if you have no money."

In fact, a constructive version of "stay at home" was the most frequently cited advice among 100 tips offered by freelance members of the Society of American Travel Writers surveyed for this text.[17] This idea—write about where you live as a "destination" to others—was discussed in Chapter 4.

Other good advice came forth, as well. Here's a sampling:

- Eat like a local—Book hotels where the price includes breakfast. Make breakfast the day's main meal. Take fruit, yogurt, and bread from the dining room after breakfast—that's lunch or a late-afternoon snack. Shop a local market for dinner.
- Getting around—"Cover [the] destination on foot as much as possible," one freelancer advises. "Not only money-saving but you see more and hear more." Same for transport: subways and busses, not taxis. "It's cheap and makes a better story," another freelancer says.
- One trip, many stories, more than one market—With planning, it's possible to "get lots of stories out of one trip," one freelancer advises. "Then rework the information to suit lots of different [outlets] so you can sell the story multiple times."

- Another job as a sinecure—This is one step up from "stay at home." Now, it's "start with a part-time job that allows you to pay the rent and eat while building a career in travel writing," one freelancer recommends. If abroad, teaching English as a second language is a good approach, as discussed in Chapter 7.
- Advance research—Don't spend time (and money) while traveling to gather information that's available at home. Both academic and public libraries still employ reference librarians. Use them to guide you to secondary sources, data, and documents. Between Google Earth, Skype, and email there are low-cost ways to see, hear, and connect with a destination long before you arrive. Don't let advance research get in the way of serendipity. But don't travel in search of information available (at much lower cost) in advance.
- Credit cards, receipts, and accounting—Several freelancers recommend putting all expenses on a single credit card known for its travel-related rewards—either miles or points. "I can fly free at least once a year," said one. Ask an accountant for advice about the tax deductibility of expenses. If your travel journalism qualifies as a "business," yes. If a "hobby," no. Profitability is the key to yes or no. Keep receipts. Log mileage.

## Saving Money—and Finding Stories—on Accommodation

One expense—accommodation—deserves more attention. Accommodation is both a problem and an opportunity. The problem arises from the high cost, especially in the major urban centers. The opportunity is journalistic: Make the search for accommodation—affordable, chic, accessible, kooky, adventurous, whatever—an aspect of the travel journalism portfolio.

Maybe it's a News-type story about the growth of "pod" hotels in cities such as Tokyo, Amsterdam, and New York. Or a Service and Advice-type story about European hostels that target families. Or a Destination-type story about WWOOFs (World Wide Opportunities on Organic Farms)—that is, organic farms that exchange accommodation for work. Or maybe it's a Journey-type story about, well, finding accommodation.

Here's background on a range of options, based on developments through 2011:

- Pod hotels—Major urban centers are both a problem and an opportunity. The problem arises from inexperience: It's hard to know where or how to book accommodation in a first visit to London, Paris, or Rome. Feeling uncertain? Book according to the advice in an up-to-date, mainstream guidebook. For the more adventurous, some of the most interesting developments in accommodation are occurring in major urban centers. It's been five years since *Fodor's* first took notice of this trend with the article "Pod Hotels: Small, Stylish, and Cheap."[18] By "small" they mean as small as 55 square feet, enough room for a double bed and a foot or so of walking space on two sides. No desk, no chair, no set of drawers. A flat screen TV is flush against one wall. Cleaning, replacement linens, and towels are extra. But the rooms are clean, air conditioned, and quiet. Brand names include the fast-expanding easyHotels, as well as Pod Hotels, Qbic, Yotel, and CitizenM. Rates are as little as $35 per night. An earlier, similar accommodation is the capsule hotel, common in Japan but not elsewhere. Yotel is the closest in design.
- Couchsurfing—Some three million people from 235 countries have found sleeping space on someone's couch through exchanges such as couchsurfing.com.[19] Couchsurfing attracts a younger demographic: The average couchsurfing.com member age is 26.[20] Airbnb is now making it a business. In more than 8,000 cities, Airbnb "rents houses, castles, cars, yachts—even igloos," as well as lowest-cost floor space.[21]
- Homestays—A cultural experience as much as an accommodation, homestays are both urban and rural. Homestays—also referred to as "hospitality stays"—are based on the concept of exchange. "When you are a traveler, you stay with others," writes Tim Leffel. "When you are stationary, you volunteer to be a host."[22] Stays are arranged through organizations. Servas is the largest, with 14,000 hosts in 130 countries.[23] There are dozens of other homestay organizations, most organized by geography. Homestays are not bed and breakfast: Conversation, sharing chores, exploring together are common. That's the point, and the journalistic opportunity: cultural literacy. Travel journalist Kari J. Bodnarchuk recounts staying with a woman in Jakarta. "By the time I left Margaretha's home, three days later," Bodnarchuk writes, "I understood more about Indonesian culture than I had learned during two months of travel in the country. And I'd made a friend for life."[24]
- Farmstays—The exchange here is accommodation on a working farm or ranch in return for light farm labor. Best known among sponsoring

organizations is WWOOF (World Wide Opportunities on Organic Farms), now 40 year old.[25] Labor might include pruning vines, milking cows, or feeding livestock.[26] Accommodation varies from luxurious en-suite bedrooms to bunkhouses. Farmstays are most common in Europe, Australia, and New Zealand. As with homestays, cultural exchange underpins farmstays.

• House-sitting—Home-sitting isn't new, but the online exchange Andrew Peck is building is. Trustedhousesitters.com matches house-sitters with house owners. The sitter takes care of property and pets in exchange for free accommodation. "We need a house-sitter for our home for at least three months," writes a couple in Kuala Lumpur, Malaysia. "We are a newly married couple and would be going on our honeymoon as soon as we conclude this arrangement." Sittings range from a couple of weeks to three months or longer, across five continents. Writers are common among house-sitters, according to Peck, perhaps exploiting the opportunity for immersion in a place. Some house-sitters are paid.

## Gadgets and Apps Every Travel Journalist Should Carry

Among the dozens and dozens of popular, travel-related gadgets and apps, here are four:

Google Translate—A free app for smartphones that supports nearly 60 languages. It displays translations in the local alphabet—perfect to show bellman, taxi drivers, porters, or police. Press a button and a native speaker voices the translation.

Dropbox—A free app for on-the-road backup of any file—text, photographs, video. Any file placed in Dropbox on a Mac or PC is automatically backed up and accessible to any of your other computers. The 2GB service is free; it's $99 a year for up to 50GB.     ·

Tripit—A free app that organizes travel itineraries. Forward emailed itineraries to Tripit and the app organizes the files, keeps track, and notifies you when anything changes.

Battery backup—Two smart choices for travelers caught short with a low battery and no place to recharge. Both Stitchway and Monoprice offer cheap, lightweight blocks that will double battery life.